Spatial Complexity in Urban Design Research

This book offers state-of-the-art 'tools for thinking' for urban designers, planners and decision-makers.

Thematically it focuses on the contexts of problems in urban design and places community spaces at the heart of urban design research. The book provides practicable tools for network modelling and visualization in urban design research. Step-by-step examples take readers through methods for tracing the evolution of road networks, and their impacts on contemporary community spaces. Easy-to-follow guides to programming show how to process and plot community data sets as network graphs. They reveal how these can help to observe and represent the different ways in which community spaces are inter-connected. This book places these technological methods in the context of current theories of community formations. It considers how these cutting-edge tools for thinking in urban design research – comprising both theories and methods – could transform our understanding of community spaces as being complex, inter-dependent and socially meaningful assets.

This book is pioneering in its analysis of the urban contexts to community formations, and in its argument for professional integration between urban and knowledge practitioners. Academics and professionals within the fields of design research, urban studies, spatial analysis, urban geography and sociology will benefit from reading this book.

Jamie O'Brien is a design research specialist at URBED, working with spatial data analysis and visualization. He was Research Fellow at Salford University's School of the Built Environment, and Senior Research Associate at the Bartlett Centre for Advanced Spatial Analysis, University College London (UCL). He holds a EPSRC doctorate in design from the Bartlett School of Architecture, UCL, and is Fellow of the Royal Geographical Society, member of the Design Research Society, and Fellow of the Higher Education Academy.

Routledge Critical Studies in Urbanism and the City

This series offers a forum for cutting-edge and original research that explores different aspects of the city. Titles within this series critically engage with, question and challenge contemporary theory and concepts to extend current debates and pave the way for new critical perspectives on the city. This series explores a range of social, political, economic, cultural and spatial concepts, offering innovative and vibrant contributions, international perspectives and interdisciplinary engagements with the city from across the social sciences and humanities.

Art and the City
Worlding the Discussion through a Critical Artscape
Edited by Julie Ren and Jason Luger

Gentrification as a Global Strategy
Neil Smith and Beyond
Edited by Abel Albet and Núria Benach

Gender and Gentrification
Winifred Curran

Socially Engaged Art and the Neoliberal City
Cecilie Sachs Olsen

Peri-Urban China
Land Use, Growth and Integrated Urban–Rural Development
Li Tian

Spatial Complexity in Urban Design Research
Graph Visualization Tools for Communities and their Contexts
Jamie O'Brien

For more information about this series, please visit www.routledge.com/Routledge-Critical-Studies-in-Urbanism-and-the-City/book-series/RSCUC

Spatial Complexity in Urban Design Research

Graph Visualization Tools for Communities and their Contexts

Jamie O'Brien

LONDON AND NEW YORK

First published 2019 by Routledge

2 Park Square, Milton Park, Abingdon, Oxon, OX14 4RN
605 Third Avenue, New York, NY 10017

Routledge is an imprint of the Taylor & Francis Group, an informa business

First issued in paperback 2020

British Library Cataloguing-in-Publication Data
A catalogue record for this book is available from the British Library

Library of Congress Cataloging-in-Publication Data
A catalog record has been requested for this book

ISBN: 978-1-138-65205-7 (hbk)
ISBN: 978-0-367-72826-7 (pbk)

Typeset in Times New Roman
by Apex CoVantage, LLC

For Lynsey

Contents

Acknowledgements viii
List of figures ix
List of tables xii

1 Design research in the built environment 1

2 Design thinking and spatial complexity 11

3 A case-based approach to design research 23

4 Building on the past 36

5 Community network toolkit I 59

6 Community network toolkit II 79

7 Building and representing knowledge 99

8 Cases in urban community formations 106

Epilogue 139

References 144
Index 152

Acknowledgements

Some of the materials covered in this book were produced during a three-year research project at University College London, which was supported by a generous Leverhulme Trust Research Project Grant. I would like to express my gratitude to the Trust for supporting this work. I also thank the Principal Investigator on the project, Prof Andrew Hudson-Smith and also Laura García Vélez and Dr Martin Zaltz Austwick at the Bartlett Centre for Advanced Spatial Analysis, University College London. I thank the project's co-investigators Prof Sophia Psarra, as well as Dr Sam Griffiths, at the Bartlett's Space Syntax Laboratory, and Prof Anthony Hunter at UCL Computer Science. I hope that this volume serves to fulfil some of the more experimental strands of the research project. Some other aspects of the work described in this book were undertaken under a CeMoRe Fellowship at the Centre for Mobilities Research at Lancaster University's visionary Department of Sociology. I am grateful to CeMoRe's director Prof Monika Büscher for awarding me this opportunity. I also pay a mark of respect to the late Prof John Urry, CeMoRe's founder, and acknowledge the legacy of his intellectual generosity. I am grateful to my colleagues at URBED (Urbanism Environment and Design), who took part in the visual logics workshop described in this book, and whose outstanding work continues to inspire my thinking around design-research problems. I owe a debt of gratitude to my wife Lynsey Hanley, whose own journalistic work continues to inspire me, and to challenge my outlook. Lynsey has encouraged my research endeavours, their successes and frustrations, not least as this book neared completion. I would not have got from there to here without Lynsey, and I dedicate this book to her.

Figures

3.1 Locations of communities sampled over socio-economic status in Liverpool, UK (based on IMD; low-to-high values shown on a dark-to-light ramp). Pie graphs displaying number and distribution of participants in each workshop site. 27
3.2 Emoticon stickers used in the Liverpool map-making workshops. 28
3.3 Urban integration cores for the Liverpool, UK, conurbation. Local scale based on 400m (left), and global scale based on 5000m (right). 30
3.4 K-nearest neighbour graphs of the Liverpool group of sampled communities. 31
3.5 Urban morphological models for Liverpool and the Mersey conurbation sample, based on integration at scales of (left) R400 and (right) R5000. The maps represent the highest affordances for origin-to-destination movements at these scales (based on the top 20% integration value range). 32
3.6 Segments at which NAIN R500 segments overlap NAIN R5000, for the Mersey conurbation sample. 33
3.7 A choropleth array representing total lengths of network segments with 'overlapping' scales of potential movements. 34
4.1 Annotated map of Liverpool's street network in the 1890s, showing areas sampled and key locations described in the study. 37
4.2 An array of Depthmap models of Liverpool, UK, representing the syntactical evolution of urban regional integration over four historical periods (NAIN Rn): 1850s, 1890s, 1950s and contemporary. Segment line widths have been graduated from high (heavy line) to low (light line). 40
4.3 An array of Depthmap models of Liverpool, UK, representing the syntactical evolution of urban regional choice over four historical periods (NACH Rn): 1850s, 1890s, 1950s and contemporary. 44
4.4 Arrays of Depthmap models of Liverpool's Princes Avenue area (1890s map). 49
4.5 Arrays of Depthmap models of Liverpool's Scotland Road area (1890s map). 51

4.6 Depthmap array showing top quintile value ranges for NACH 400 and NAIN 5000, highlighting the ways in which local high choice around Paddy's Market may have coincided with a city-wide attractor. 52
4.7 Arrays of Depthmap models of Liverpool's Canning Place area (1890s map). 54
4.8 Arrays of Depthmap models of Liverpool's Canning Place area (1970s map; detailed). 55
5.1 Simple network resulting from the graph code. 69
5.2 Geo-located network graph (metric scale shown) with iconographic vertices. 72
5.3 Network graph with edges weighted by betweenness and vertex size adjusted for community weighting. 75
5.4 Network graph with vertex sizes adjusted for closeness. 77
6.1 Plot of selected community assets over sample road network. 86
6.2 Connections among vertices fixed to geo-coordinates. 89
6.3 Cluster diagram of network graph, based on the spinglass algorithm. 91
6.4 Directed graph of relationships among selected community assets, vertex sizes adjusted for community weightings. 97
7.1 Visual representations of the RCC8 language for spatial relationships. 102
7.2 Visual representations derived from Lakoff's schema for describing spatial relationships. 103
7.3 This participant's concept graph represents a possible journey from home to a supermarket. 104
8.1 The LSOAs for the case-study sites of North Group and South Group in the broader urban context of the sampled Merseyside conurbation. 107
8.2 Index of Multiple Deprivations (IMD) for North Group spatial context (LSOA by decile, ONS 2015), where 1 is the most deprived. 109
8.3 IMD for South Group spatial context, as above. 110
8.4 Geographic barriers for LSOAs within the North Group community area. 111
8.5 Geographic barriers for LSOAs within the South Group community area. 112
8.6 Schematic map of land use (all use types except residential) for the North Group community area. 113
8.7 Schematic map of land use for the South Group community area (as in Figure 8.6). 113
8.8 Road-network integration for the South Group community area with major features selected by the two community sub-groups. 114
8.9 Road-network integration for the South Group (as in Figure 8.8). 115

8.10 Distributions of points aggregated by their 'positive' and 'negative' associations, respectively, for the North Group. 116
8.11 Distributions of points aggregated by their 'positive' and 'negative' associations, respectively, for the South Group. 117
8.12 Distributions of positive and negative sentiments associated with open spaces (especially public parks) for the North Group participants. 121
8.13 Distributions of positive and negative sentiments associated with open spaces (especially public parks) for the South Group participants. 122
8.14 Distributions of positive and negative sentiments associated with roads and road infrastructure among the North Group participants. 123
8.15 Distributions of positive and negative sentiments associated with roads and road infrastructure among the South Group participants. 124
8.16 (a–d) A bar chart array representing the counts of predominant keywords with both positive and negative associations for each of the communities sampled. 125
8.17 Assets associated with positive social encounters, for the North Group relating to keywords with positive associations. 127
8.18 Assets associated with positive social encounters for the South Group. 128
8.19 North Group community network graph with geo-located layout, plotted over street-network section. 129
8.20 North Group community network graph with force-directed layout (Kamada Kawai format). 130
8.21 North Group community network graph with vertex labels. Vertices with highest closeness centrality values are highlighted black. 131
8.22 South Group community network graph with geo-located layout, plotted over street-network section. 132
8.23 South Group community network graph with force-directed layout (Kamada Kawai format). 133
8.24 South Group community network graph with vertex labels. Vertices with highest closeness centrality values are highlighted black. 133
8.25 Directed graph of the North Group community network. 135
8.26 Force-directed graph of the South Group community network. 136

Tables

5.1 Betweenness values for the sample's network edges 75
5.2 Closeness values for the sample's network vertices 76
6.1 Manual record of connections between assets (vertex list) 87
6.2 List of weights applied to the vertices 95
8.1 (a and b) Tables of the frequencies of main features selected by the North Group community participants, arranged by age group 118
8.2 (a–c) Tables of the frequencies of main features selected by the South Group community participants, arranged by age group 119

1 Design research in the built environment

General introduction

Human societies configure all manner of technologies to deal with natural uncertainties. One such set of technologies is built from the basic urban forms of roads, open spaces and landmarks, which are arranged simply and hierarchically into towns and cities. They are simple because they follow basic, repeated patterns of conduits and concaves – of roads, squares, parks and community boundaries (Lynch, 1960). They are hierarchical in the sense that they are made up of sub-components, such as neighbourhoods or commercial zones, that feature high-frequency social and economic interactions, and that these sub-components are loosely inter-connected by certain social ties or flows of movement that transcend their spatial boundaries (Alexander, 1971). In this way different sections of the town or city, such as the metropolitan centre and suburban periphery, can be joined by commuter flows or logistics traffic.

This simple, hierarchical spatial patterning has underpinned the social and structural systems of most urban settlements. This kind of patterning has proved to be adaptable to myriad architectural and urban forms, and resilient to environmental and historical pressures and changes. A spatial pattern developed for one society can, over time, be readily adapted to the needs of a successive society. Architectures and urban forms may be erased by catastrophic events, for example by wartime bombing or major floods and fires, yet the settlement's underlying spatial pattern remains intact, immediately available as a basis of movement and rebuilding. The spatial pattern serves as the memory of the town or city, demarcating its time-honoured formal and informal relationships: its places, thresholds and boundaries.

The inherent resilience of towns and cities to environmental, historical and technological changes, based on this simple, hierarchical patterning, continues to draw people to settle *en masse* in urban areas. Towns and cities help to lengthen life and promote good health, happiness and pleasure among their inhabitants. This continuous urban revolution also provides the context for – or perhaps the cause of – social exclusion, spatial segregation, poor health and misery. Needless to say, urban settlement does not automatically lead to health and happiness. For this reason, urban designers and planners have learned how to intervene in urban developments for the benefit of their inhabitants, by understanding the ways in which forms and patterns underpin connectivities among people and places. They harness the properties of spatial complexity, stemming from simplicity and hierarchy, to promote welfare through successful social encounter and control of movement.

Urban designers and planners make use of a diverse toolkit for critical thinking, reasoning and decision-making around these themes (cf. Johnson, 2010). Many design 'tools' are embedded in working practices, and owe more to capabilities learned through studio and field work, than through technical competences attained in professional training. Both capabilities and competences play important roles in design and planning processes, and tool-use must be learned through 'guided discovery' (Ingold, 2000, p. 356), honed through practice, as well as being understood in the context of received standards. Tools held in common bind together communities of practice for design (Wenger, 1998); they serve as boundary objects for sets of know-how and, with advanced practice, for mastery.

For human societies, design tool-use becomes the means of articulating our inner social and physiological needs to achieve a 'best-fit' transformation of our common outer worlds. In this regard design tools can comprise plans and protocols as they can instruments and devices. All design tools are tools for thinking with; their use in practice shapes how the designer approaches, apprehends, conceives and imagines the design problem. Indeed, the nature of the representation of the problem is fundamental to the process by which the design is conceived as the difference between the current state of the problem, the desired state of the solution and the process by which the plan for implementation successfully transforms the problem (Simon, 1996, p. 132).

This book focuses on the potential for a set of tools for design thinking – for the representation of the problem space, and the potential plans for its transformation – based on patterns of connectivity among people and places. Our focus is those urban forms and patterns that serve to weave together the hierarchical sub-components of urban settlements, from which their adaptive resilience is founded. These social and spatial connectivities can be represented as networks, for which several well-established analytical tools exist. This book provides an overview of their application to design thinking around themes of urban transformation and resilience. It also outlines some novel methods for configuring network models over which different kinds of data can be applied for inclusion in decision-making processes.

In representing problem space designers are asserting a set of signs to represent the realities of a lived world. As designers, we should become aware and 'reflective' of the ways in which these sets of signs configure our professional vision of lived experience (cf. Schön, 1983). Towards this goal, it is worth outlining some of the critical approaches to the representation of space presented by Henri Lefebvre in *The Production of Space* (1991). Lefebvre's theory of space stemmed from a concern with the dynamics of opposing forces, including notions of interior and exterior or centre and periphery. His thinking was rooted in Marx's historical materialist dialectic, which comprehended both 'things and their representations', hence their inter-connections, origins and endings (Engels, 1884/1991, p. 390).

Lefebvre's dialectical method drew also from a notion of mirroring of what is signified (unified nature), and the effects of being signified (fragmented nature). The representation of social space as a 'game of mirrors' (ibid., 212) passes – like Alice through the looking glass – from the world of lived experience into that of reversed signs and images of that world.

In developing a unifying theory of space, Lefebvre set out to understand how our dwelling and thinking have become fragmented into many layers of cultural, political and intellectual categories (ibid., 83). Space at the global level was for Lefebvre like 'white light': it appears on the surface to be homogenous, yet it can be differentiated into a spectrum of colours under the appropriate tools of analysis (ibid., 352). Lefebvre's spatial spectrum included mental spaces representing conceptual abstractions, religious or tyrannical spaces representing notions of the absolute, spaces of our social interactions, our human desires and cultural distinctions.

While recognizing the relativity of nature, stemming from the instability of opposing forces (ibid., 65), his concern was with reversing the trend towards differentiation in social practices, and he sought to understand how nature has become separated into 'mental spaces', relating to our metric instruments for conceptualizing space; based, that is, on axes, pivots, perspectival lines and centre-points (ibid., 5–6). He sought to explain how these mental spaces represent the perspective of the dominant social class, and its intention to encode space based on a supervening logic or strategy, such as the systemic division of labour (ibid., 7).

We are not credibly able to overcome the deeper issues of fragmentation that Lefebvre outlined in his classic work. These issues stem of fundamental inequalities in the control of production and knowledge, with their origins in prehistory. However we can include reflexivity in the design process, by which we declare the fragmentary nature of our design perspective, and we avoid the desire to encompass and speak for the lived experience of our design subjects (our clients, participants, occupiers). We avoid the 'white light' of the total architectural vision, and instead illuminate the contexts in which forms take shape (Alexander, 1971, p. 91). We can use spatial analysis to describe the constraints to our creative vision, and to suggest beneficial configurations of space for sensory experience and social interaction. We can also suggest how materials and structures could serve to memorialize social meanings, whether those of the everyday street corner or those of historical events. In these kinds of practice, our reflexivity relates to an acknowledgement that space is a spectrum, it is the product of a convergence of multiple geometries, scales and practices that flow through its form and its context.

Concepts for designing

Urban planners have applied renewed emphasis on harnessing citizens' place-specific experiences for successful project implementations (cf. UN-Habitat, 2009, p. xxiii; Cassiers and Kesteloot, 2012; Berman, 2017). Community collaboration as part of a heuristic, multi-dimensional model (including stakeholders, citizens and planners) has been cited as a crucial component for effective planning policy interventions, across a series of purposes (Innes and Booher, 2004): decision-makers need to find out about the public's preferences in the planning domains, they must incorporate local knowledge into the domain parameters, including those of the least advantaged groups (which are not revealed in the standard analysis or privileged viewpoint). Participation may also help to secure legitimacy for a planning decision (albeit this could lead to 'box-ticking' on the part of the planning officers) and build trust among stakeholders and citizens.

More generally, participation in public policy and planning domains has been recognized as a driver of a 'more intelligent society' (Innes and Booher, 2004, p. 431) that is better able to adapt to changing or difficult conditions. Yet current professional practices and norms often militate against collaborative participation. *Professionalism* imposes generalized standards of practice on urban initiatives, often circumventing locally available knowledge that would support bespoke, sustainable solutions (Childers et al., 2014; RTPI, 2005). *Distance* results from centralization, both in terms of geography and administration, which similarly leads to professionals' poor understanding of local contexts. *Power* hampers meaningful dialogues, shared understanding and organizational learning, through which local knowledge may be generated and exchanged (Chambers, 1997, pp. 31–21).

In order to address the problems of professional practices in local community contexts, it has been widely recognized that community participants must be empowered to articulate their positions in terms of their localized definitions (Chambers, 1995), and to articulate their social connectedness and access to economic opportunities (cf. Morsey, 2012; United Nations, 2013, p. 77). Current information technologies have enriched urban-domain process designs (Ramasubramanian, 2010, pp. 34–37); however, geographic technologies also pose challenges of uptake and use in professional and scientific domains (Tulloch, 2007), as well as to the diversity of data producers, in terms of their expertise and motivations for participating (Coleman, Georgiadou and Labonte, 2009).

A further challenge relates to the embeddedness of participants' knowledge. Community members' concepts of their 'neighbourhood' may, for example, vary according to their age, gender, level of ability, socio-economic standing or stage in life (Lupton, 2003). For these reasons and others, we find an epistemological split between professional and 'inhabitant' spatial cognitions (Ingold, 2011a, pp. 146–155). Stemming from this split, various urban sociologies have described ways in which professional observers overlook supposedly 'trivial' streets that inter-connect blocks of tertiary streets (Grannis, 1998), thus providing a path for novel community links to form. Elsewhere, officially designated infrastructural projects have been observed to undermine or replace 'community anchors', such as places for stopping and chatting or for children's play (Power, 2007, pp. 58–59).

Problems of complexity and inequality

Here we draw from Lefebvre's notion of the urban inhabitant as organism and its extensions into space, without which it has 'neither meaning nor existence' (Lefebvre, 1991, p. 196). Lefebvre's dialectic of urban space focuses in the inherent contradictions brought about from this notion. Firstly, those contradictions of the built environment as a hyper-complex expression of its inhabitants' productive and reproductive forces (c.f. ibid., pp. 87–88). Secondly, those contradictions of the built environment as a super-positional form, whereby signs of the dominant or historical spatial orders are imposed upon subaltern or contemporary landscapes (ibid, p. 55). The inherent contradictions of these forces are born through the subjects' biological and socio-cultural struggles and conflicts, through which the ideology of the dominant order – expressed in geometric codes of the urban landscape – becomes erased by everyday lived experience of the inhabitant's body in space (ibid, p. 40). For this reason, the supposed 'legibility' of the well-designed city (c.f. Lynch, 1960, p. 2), may be configured for the ease of navigation among the dominant group. The city is shaped arguably around 'their' knowledge of its spatial logic; it represents their class interests, formalized through patterns of movement, visibility and access.

Lefebvre drew on the analogy of the automotive driver, who fragments urban spaces into a 'lived abstraction' of signs for driving, known only to himself (the sign-bearing 'I') and to other 'sign-bearers'. The driver reduces the total knowledge of the built environment to streams of information, to the 'materialized, mechanized, technicized' route (Lefebvre, 1991, p. 313). While we should not get lost in this analogy (automotive driving is no longer the preserve of the dominant group in the West), we may similarly consider the impact of urban systems (mobilities systems, informational systems and so on) on knowledge of the built environment. Both our group actions (as subsets of the total population) and our particular modes of engagement with the environment draw from its wholeness those informational sections that instrument or utilize our desires. There are many ways in which we reduce our embodied knowledge to the commands of an informational sign-bearer.

We focus on the pertinence of these themes to urban planning and design processes. We seek to understand how the contradictions of built-environment domains are acknowledged in these processes, including their role in the applications of geographic information systems (GIS) in participatory engagements. There is increasing recognition among institutions of affluent, industrialized ('Western') contexts over the need for community participation in these processes. In these cases, the value of iterative, participatory practices has been located in the need for sustainability, resilience, prosperity, minority inclusion, community cohesion and conflict prevention (Roggema, 2017). Dialogue among community members and municipal officers has been shown to promote the successful implementation of the urban project (UN-Habitat, 2016, p. 10), while community engagement has been shown to enhance the post-occupancy sense of public ownership (Hay, 2011, pp. 9–12).

Participation and innovation

Urban practitioners, and urban planners in particular, may face a series of dilemmas in their professional practice, stemming from the tensions among community and professional perspectives over what constitutes the 'common good' or 'public interest' (Ramasubramanian, 2010, pp. 10–12). In the 1960s, the advent of computational methods in planning brought about top-down, system-based planning strategies for many American and European cities; the negative effects have been seen in radial highway initiatives that connect regions to metropolitan centres, but fragment and impair the environments of local community spaces. Digital technologies were used in these instances to enframe a professional vision that excluded the requirements of daily living at the community level. However, digital technologies have also driven a rebalancing in power relationships among professional and community practitioners. Geographic information or environmental sensor systems may be used in conjunction with 'hand-made' activities to gather, manage and make use of rich data and documentation pertaining to participants' everyday experiences and requirements. Such adoptions of technologies have transformed the community into 'enacting organizations', meaning they are proactive, interpretative and adaptive to the changing environment of innovations (Seely Brown and Duguid, 1991). The organization also brings about a shift in knowledge attainment away from the centre (for example from the viewpoint of the professional planner) to the integrative periphery, to produce what has been termed the 'peripheral legitimacy' of communities of learning (Lave and Wenger, 1991). This shift in the production and attainment of knowledge has parallels with Eric von Hippel's influential work on the networked nature of innovation, whereby technology participants at the 'periphery' of the network make enhancements to the innovation not initially envisaged by the innovator, forming a 'value network' of shared knowledge assets (von Hippel, 2005; Vanhaverbeke and Cloodt, 2006).

Design thinking and reflexivity

The notion of a value network of non-professional participants and professional practitioners may pose a paradigmatic challenge to so-called 'design knowledge'. This kind of knowledge is embedded in professional know-how, produced through active engagement with given problems. Urban practitioners engage their 'design knowledge' in the project process and implementation, which involves a combination of deductive experimentation, practical know-how and the sense of 'what should be'. They bring to the problem space sets of professional and personal values and 'templates' for configuring possible solutions, often based on experiences from other project domains. Design pedagogy is often concerned with the students' practical engagement with given problems and overcoming difficulties in understanding long-established professional conceptualizations (parenthesized, or 'chunked', into technical jargon), to achieve engagement with the 'common ground' of canonical designs, through which widely adopted design patterns are established, and to engage in critical or peer reflection on their design iteration (Hoadly and Cox, 2009).

Design activities have been observed traditionally as belonging to either of two frames: those of rational problem-solving and those of reflective practice (Doorst and Dijkhuis, 1995). The 'rational' frame extends from a notion that the designer is an information processor within an objective reality, seeking optimal results from poorly structured problems. The 'reflective' frame extends from a notion that the designer constructs his or her own problem situation through dialogue and creativity. Design thinking by professional and non-professional practitioners alike arguably includes both approaches. Our approach to a problem is rational in the sense we intuitively distinguish its stable forms and functional patterns, and then seek the optimal opportunity to make a change. It is reflective in the sense that we pause to observe what our actions produced, and to discuss this with those who are also involved in the problem. The solution may not be computationally optimal, but it becomes the 'best fit' for the material, functional and social constraints of the problem situation.

Urban environments as highly complex and relational artefacts – involving people, places, structures and protocols – pose so-called 'wicked problems' to the urban practitioner. The requirements and parameters of the problem space may be highly difficult to analyze due to the complexity or repetitiveness of components within the system, or due to factors lying outside of the known design space (Rittel and Webber, 1973). In short, the so-called wickedness of these problems stems from their complexity, recursion and obliqueness. In urban domains, our design practice can relate to artificial systems (based on computing and robotics, etc.) and also social systems, as well as to combinations of artificial and social systems. This poses further dimensions of human complexities to 'wicked problems', whereupon the interests and perspectives of various individuals and groups contribute to the problem at hand. Troublesome knowledge has been described as 'knowledge that is "alien", counterintuitive or even intellectually absurd at face value' (Meyer and Land, 2006, p. 4). Wicked problems by their nature present both the professional and non-professional practitioners with the challenges of reasoning and understanding. One way of overcoming these challenges involves practitioners delving into problems, to experience first-hand the many dimensions of the problem, and to draw upon their breadth of lived, creative and technical experience in producing sets of requirements and solutions to the problem.

This delving into the problem situation requires practitioners to engage in 'authentic dialogues' with co-inhabitants of the situation. The primary purpose of engaging in dialogues is to drive collaborative participation, rather than to build simply a 'pluralist' packing together of individual interests (Innes and Booher, 2004). Yet to achieve authenticity is to suspend much of the parenthetical jargon of specialist higher education by which we *think of* urban spaces. The intermediaries of urban community spaces – the formal spatial properties of their boundaries, thresholds and interfaces (Palaiologou and Vaughan, 2012) – provide the means by which urban participants *think with* their environments (Hillier, 2007, pp. 27–30), in terms of everyday activities based upon spatially embedded meanings.

We maintain a position here that *thinking of* and *thinking with* urban spaces each requires dialogue-based interactions between actors and their environments. *Thinking of* involves contrasting definitions of urban spaces, which are exchanged discursively among actors, and which compel iterative refinements and calibrations in practitioners' designs and models. *Thinking with* urban spaces are communicated non-discursively through sub-conscious exercises, yet compel dialogic exchanges of spatially embedded meanings among actors.

Overview of the book

This book takes on two main tasks. The first task it to outline the themes of reflexive design thinking and what kinds of design-thinking tools might support this. This task will be undertaken through a discursive overview of themes in critical design and planning literature, which we have covered in this introductory chapter and continue this discussion in Chapter 2. In Chapter 3 we outline a case-based approach to the issues set out in the book, which is based on some earlier research into community formations among adolescents in Merseyside, UK. Chapter 4 describes another section of this project that sought to understand the impact of the historical urban network on contemporary communities. This work was based on an application of space syntax to road-network analysis, which provides a 'normative' approach to the analysis of network integration and is useful for comparing networks over time. Reflections on this overview and approach to design thinking will be presented in the concluding chapter.

The second task is to present the components of a toolkit to support reflexive design thinking, which is based on the 'graph data frame' structure that is included in the open-source iGraph programming library. This structure is highly flexible in general and spatial data analysis, allowing the programmer to attach any kind of data and data structure, including the standard structures of lists, arrays, matrices, tables, data frames and rasters. The 'graph' component of the data frame allows the programmer to represent nodes with single or multiple layers of information and, crucially, to represent connections between these nodes in the forms of edges. Edges can themselves hold data including any mean values that relate to their adjacent nodes. In short, graph data frames allow us to represent real things in space and the relationships between them. They help us to approach the subjects of our design thinking as complex or composite entities, shaped through the spectrum of their social worlds.

The tool has been written in the R programming language that is, arguably, a de facto open-source tool for data analysis. R's wide adoption in this field is possibly based on its set of compiled functions for processing data (the so-called 'ply' family). We do not go into detail about these functions in this book, and instead make use of the 'for loop' function for data iteration. This potentially introduces some computational costs, albeit for processing data sets far bigger than those described in this book. Readers are welcome to compile and run the R code,[1] and investigate any alternative approaches to the programming tasks. Readers not interested in the programming methods may also skip over much of the contents of these chapters and still enjoy the book's main themes.

The examples are drawn from a study that formed part of a broader project at University College London called 'Visualizing Community Inequalities', which ran from 2014–2017 and was supported by a Leverhulme Trust Research Project Grant. The details of this section of the study are presented in Chapter 3, and in

Chapter 8 we attempt to describe ways in which the communities sampled have established their community spaces. Specifically, we apply our tools to describe the composite nature of boundary markers used by the participants. This is a crucial matter for design and planning around urban community contexts, as these boundaries and their markers are often invisible to the external observer. They exist compositionally as physical and conceptual artefacts that relate to sets of meanings embedded in the community context. For example, we cannot 'see' that a public park contains a threshold somewhere along an ostensibly open field, across which has been set the limits of two or more communities in the area. Similarly, certain streets, junctions, landmarks and other spatial structures demarcate a community boundary beyond which its members feel less safe and secure, or indeed directly threatened. Finding some ways to describe these kinds of 'relational artefacts' (Hillier, 2007, pp. 67–68) in their spatial contexts is the central purpose of this book. In the concluding chapter we attempt a synthesis of the book's concerns and themes, and suggest possible avenues for further work in this field.

Structure of the book

Following from this general introduction, this book develops a line of argument about the potential for graphic tools in design thinking, especially as it forms a critical relationship with the artefacts of urban community formations. In Chapter 2 we consider the importance of design thinking in achieving sustainable urbanism, stemming from the representation of community spaces in terms of their inter-dependencies. In Chapter 3 we introduce our case study in urban community formations, and some of the technological methods we employ to illuminate their historical and contemporaneous spatial contexts. Chapter 4 describes the historical development of the road network for the case-study context. This historical description is important in understanding the impact of infrastructures on contemporary formations, especially in the distinctions between dense, rectilinear, or 'grid-pattern', street layouts of the 19th century period and the lower-density, 'concave' layouts of the 20th century. The evolutionary nature of the road network is also described in terms of incremental consolidations of ancient or 'accidental' pathways, and of the possibilities for recovery of architecturally subsumed pathways.

Chapters 5 and 6 take a turn towards technological development of a toolkit for describing community formations based on network graphs. Firstly, we outline a method for generating undirected graphs, and then for applying weights to graph vertices derived from community participatory contexts. Chapter 7 develops some experimental schemata for the iconographic representation of spatial relationships that underpin community formations in their locational contexts, such as 'externally connected' or 'overlapping', based on the region connection calculus (RCC8) language (Randell, Cui and Cohn, 1992), with additions from Lakoff's (1987) schemata for spatial relationships. Chapter 8 provides a set of worked examples in applications of the design-thinking tools that build on the themes and technical aspects of the previous chapters. This chapter presents a series of visual maps and network models by which we can observe some patterns in community relationships. In Chapter 9 we offer some concluding remarks about the potential impact of graphic approaches to design thinking on sustainable urbanism.

Data sets and applications

Design researchers, and those engaged in the illumination of contexts for problems in design and planning, can make use of freely available, free-to-use computational and data management tools, and access detailed and rich data sets. Most are freely available to the general public. For example, most of the maps and models presented in this book were generated using QGIS and the RStudio programming environment, including the iGraph programming library, which are both free to download and use. Much of the spatial data was derived from either the UK's Ordnance Survey or Office of National Statistics digital repositories. These are publicly accessible via the World Wide Web. Some of the iconographic content was developed using the GIMP graphics application, which is also free to download and use. The general availability of these applications and data sets is potentially transformative to design thinking in urban domain contexts.

Note

1 Available to download at www.routledge.com/9781138652057

2 Design thinking and spatial complexity

Practitioners in urban design and planning – professional and non-professional alike – engage in iterative, problem-orientated 'design thinking' (cf. Schön, 1983; Rowe, 1987). This kind of thinking makes use of a widely adopted notion of 'affordances' in the lived environment (cf. Gibson, 1979; Norman, 1988; Marcus, Giusti and Barthel, 2016). Affordance refers to any sort of material or dynamic property of the environment that provides its inhabitants with the means to live successfully. Architectural or urban (i.e. 'built') environments provide affordances that have been shaped with some deliberation to meet the communal requirements of their inhabitants (cf. Lynch, 1960; Alexander et al., 1977; Hillier and Hanson, 1984; Brand, 1997).

However, affordance is not an evenly distributed property of the built environment. We can also observe how social and economic inequalities are reflected in highly contrasting qualities in the built environment, in terms of both fabric and functionality. For example, the affordances of the road network might provide commuters with direct accessibility from suburbs to economic centres; yet these city-scale road complexes might bypass or disrupt the affordances for localized movements among specific communities. Dealing with the inequalities of urban community spaces is of critical concern to promoting sustainable urbanism, as it relates to social, ecological and systemic factors in the provision of services. For this reason, design thinking that seeks to address spatial challenges that stem from socio-economic inequalities must challenge or negotiate the placing of wider-scale spatial assets in local contexts.

Designers can observe how communities have responded to these kinds of spatial inequalities. For example, where a regular street market has become established in the shelter of major road or railway bridges. In this instance, the bridges may have been built to afford city-scale transit, yet 'design thinking' can challenge their given functionality to produce affordances of shelter, economic activity and place-making. In this respect, 'affordance' is a latent, potential or complimentary property of the built environment, which depends for its realization on a set of problem-orientated activities embedded in the challenges of everyday living (cf. Thompson, 2013).

A topical focus on sustainable urbanism through local interactions lends natural weight to the recovery of pedestrian infrastructures, and their intersections with other modes and scales of movement. Affordances for pedestrianism can be observed as 'latent' properties of the built environment, not least where the contemporary road network has been shaped around ancient movements (Hoskins, 1955/2013, pp. 63–68). Furthermore, current technological advances in electrical and driverless vehicles impel design thinking around the alleviation of roads from private, automotive traffic and the burgeoning potential for mobilities-based services that are connected tangibly with community life (Heinrichs, 2016).

In other contexts, affordances for pedestrianism might be 'accidental' properties stemming from historical interventions. For example, we find instances of roads that have been permanently cordoned for purposes of crime prevention or community severance, with the intended effect of limiting traffic and the unintended effect of promoting street-level interactions. Developing a toolkit around the relational complexities of design affordances, as reflecting in latent properties and accidental opportunities for intervention (Hillier, 2007, pp. 67–68; O'Brien and Psarra, 2015), would be a powerful driver in the advancement of sustainable urbanism.

Design research in the built environment

Localized urban community spaces have been, more often than not, already shaped around activities of everyday living and normative processes over time (cf. Rodger and Sweet, 2008; Hillier and Hanson, 1984, p. 22). Their spatial structures – for example their roads and open spaces – have been formed through social relationships and economic movements (Pooley, 2000; Lawton, 1979; Dennis, 2008). The functionalities of urban spaces continue to stem from people's needs to remain close or apart from one another (Marcuse, 2002; Massey, 1994, p. 147; Sibley, 1995, pp. 91–92). Urban community spaces also reflect the place-specific demands and requirements of local sub-groups (Lupton, 2003; Grannis, 2005, 2009; Kearns and Parkinson, 2001), which give rise to complexities of architectural structure and geographic scale (Gans, 2002, 2006; Logan, 2012; Hanson and Hillier, 1987; Smith, 2010), and to affordances and representations of accessibility or limitation (Hillier, 1999; Hillier and Iida, 2005; O'Brien and Psarra, 2015).

In responding to these challenges of spatial complexity, some current open source technologies can be configured to support practitioners in gathering detailed local knowledge within well-defined research frameworks (Haklay et al., 2014). Advances in GIS, geographic positioning systems (GPS), digital photography and imaging and communications platforms have been made available to allow community groups to record, collate and publish their interactions in spatial contexts. These can potentially be used to challenge the assumptions of key decision-makers in contested urban domains (ibid.).

A key impact of the utility of these tools on urban design research can be seen in the redefinition in practice of the dual nature – physical and conceptual – of artefacts that comprise local urban communities. These artefacts can now be observed as being products of complex relationships between discursive and non-discursive agencies within their urban contexts.

We focus on the problems of defining conceptual artefacts by considering how urban communities' social meanings are embedded in their spatial configurations, conceptualizations and practices. Considering the relational nature of the built environment, we can describe the interplays of space, society and meaning as being 'dialogic'. This means that the urban environment's discursive and non-discursive agencies inform and transform each other through processes of their complex inter-dependencies. These dialogic processes also occur where professional and community practitioners seek to transform the built environment by exchanging their conceptualizations and definitions.

Towards a refocusing upon conceptual artefacts in the built environment, we review in this chapter a selection of diverse research from the fields of space syntax, actor-network theory in architecture and urban sociologies of crime and deprivation. We sample from specific studies of urban spatial effects upon local community behaviours. We observe that processes of conceptualization are revealed in professionals' definitions of urban environments. Moreover, we draw attention to the lack of community-membership definitions in many urban interventions. We argue that this lack persists because community conceptualizations, based upon 'mental models', tend to be reflected in quotidian or sub-conscious practices, which do not enter the standard professional discourse.

Urban practitioners have acknowledged that society cannot be reduced to space (cf. Gans, 2002, 2006). Urban communities involve subtle and irresolvable interplays of social meanings and spatial structures, forming their physical and symbolic boundaries (Logan, 2012). Community spaces also include effects from within the broader urban network (Sampson, Morenoff and Gannon-Rowley, 2002; Hillier and Vaughan, 2007), positioned in one space with multiple layers of spatial, social and effective properties (Grannis, 2009). In this way, urban community spaces have social meanings embedded in their configurations, conceptualizations and practices. Understanding this 'super-positionality' (ibid., p. 17–18) of urban community configurations, such as those found in neighbourhoods, warrants a methodology that draws upon distinctive and complementary perspectives.

In order to address these themes, in this chapter we also seek to reflect critically on a sample of urban research studies, considering, in particular, their treatment of the role of conceptualizations in shaping the urban environment. Building on this critique, we argue that the notion of 'mental models' is overlooked in the design research literature and warrants further investigation. Working towards a synthesis of physical and conceptual artefacts, we attempt an outline of the significance of inter-dependencies in urban formations. Hence, we consider the role played by local conceptualizations in phenomena such as neighbourhood boundaries, community foci, stereotypes of others and place-specific community values. Finally, we outline the requirements for a method to examine these conceptualizations.

Considering together the diverse studies we have sampled, they tell variously of what we call the *dialogic city*: an urban environment formed of inter-dependent spatial configurations, conceptualized realities and situated practices. Moreover, these components possess the capabilities to inform and transform each other. For this reason, we have argued that the notion of agency in the built environment helps us to study the relational complexities of community spatial configurations, such as those of the neighbourhood. In later chapters, we seek to develop methods that might encapsulate the spatial and semantic combinations that underpin community formations.

In the following sections, we offer a brief introduction to current work in this field. We offer an overview of disciplinary perspectives on the key theoretical themes relating to this work. Subsequently we provide a selective review relating to urban configurations, conceptualizations and practices. A discussion of this review attempts a methodological synthesis of physical and conceptual artefacts.

Perspectives on urban spaces

Agencies and effects

Community spaces are special features of the urban environment, formed through the socio-spatial configurations by which people achieve 'nearness' at many levels of the home, street and public space. Community spaces comprise relational complexes of object and abstract artefacts (Hillier, 2007, pp. 67–68), which we term physical and conceptual artefacts, respectively. By extension to this argument, people and places have agency, that is, they have quasi-autonomous abilities to shape their environments. Spatial agencies are any components of the urban community space which actively change the relationships and forms or flows of that environment.

Design interventions in community spaces, whether planned or embedded, involve networks of citizen and professional practitioners, who (together or apart) influence the trajectories of their urban projects. These practice-based agencies are discursive, in the sense that they are based upon verbal communications, and they may be talked about directly. For example, in the context of architecture, Adrian Forty has outlined the relationships between the architect's professional vocabulary and the production of meanings of space for design practice (Forty, 2004). In contrast to this view, Hillier defines space and social relations as being fundamentally non-discursive (cf. Hillier, 2007).

Non-discursive agencies are not verbal and are not talked about directly; they form conceptual intermediaries of community life. Such agencies may not be recognized consciously by professional and citizen practitioners, and necessitate analytical methods to reveal their significance and meanings within the urban environment (Hillier, 2007; Psarra, 2009; cf. Rappaport, 1990). For example, certain network configurations can underpin everyday pathways and routes, yet the configurations' basic structures are subsumed within the urban environment. So, too, geographic, political or even administrative features (such as postcodes) may tacitly demarcate significant community spaces.

One such non-discursive agency may relate to the 'neighbourhood effect' – being a causal property of the local environment with a predictable impact upon community behaviours. Research that seeks to determine these effects has been criticized for reductively defining 'neighbourhoods' based upon administrative units (cf. Lupton, 2003). Yet, in spite of these criticisms, there appear to be correlations between concentrations of poverty and frequent disorders and disadvantages in local urban community areas (Sampson et al., 2002). The specific dynamics of these kinds of effects in socio-spatial contexts warrant further investigation.

Similarly, in the area of social network research it has been suggested that community spaces may produce predictable positive outcomes. For example, social factors that mitigate disorders affecting children and adolescents[1] seem to be based, perhaps counter-intuitively, upon *weak* social ties within and across culturally homophilous communities (Grannis, 2009, pp. 25–26). Such weak ties mean that adults may know each other indirectly through their children, as well as through interactions at local community facilities, such as schools or health centres. Weak ties mean that adults are more likely to intervene in each other's children's behaviours in the street, and also to reproduce positive behavioural and social norms (ibid.). Hence, we can, for example, observe that a neighbourhood 'effect' is embedded within the hierarchical interactions of the complex community network. The studies in neighbourhood and social effects that we have sampled in this chapter have, in different ways, incorporated socio-spatial complexity in their analyses.

Impacts of tools in design

The sociological field of actor-network theory has considered the so-called 'mutual constitution' of practices and materialities. These include human and non-human actors in the process: the activities that people do, and what they do these activities with. For example, Farías and Bender (2010) have described how cities include 'non-human ecologies', which generate urban spaces through the mass interactions of their machine components. Elsewhere, urban geographers influenced by actor-network approaches have described the 'symbiosphere' of cities, comprising the inextricable networks of people and tools to produce the urban ecosystem (Amin and Thrift, 2002). Some actor-network theorists have argued that much urban research has 'hidden' cities behind disciplinary concepts such as structure, system and scale (cf. Smith, 2010).

The 'actor network' of spatial analysis is represented in multi-layered GIS integrations. These show manifold inter-relationships of populations, their neighbourhoods, the trajectories of their movements, the conditions of their socio-economic behaviours and cultural and political experiences. Yet, like all tools for visualization and mapping, they also impose artificial conceptualizations upon practitioners' thinking about the urban landscape. For example, the notion of convex space in an urban area discretizes an open space into units, while in empirical and conceptual terms these areas are perceived differently, so that no matter how divisible into constituent parts, they retain an identity as 'wholes'. Therefore an outstanding challenge in the analysis of conceptual artefacts is to distinguish community concepts from professional concepts.

Urban practitioners must, as a consequence, reflect on how their professional tools affect their analysis and descriptions of space. Reflexivity in practice may be achieved by examining how professional tools become agents in shaping the urban landscape. Professional tools may be physical or conceptual but, either way, they have 'obdurate' properties: they impose, for example, technological standards or the preconceptions of mental models onto urban developments (Hommels, 2010). The following sections build upon this observation to consider how the built environment as a whole is formed through agencies and intermediaries, including concepts, definitions and dialogues. Throughout this chapter, these themes are related to the formation of community spaces.

We also noted above how we have focused our current research on areas that present deprivations and poverty. This is because differences in conceptualization and definition appear to be particularly wide between community and professional practitioners. In the following sections, we review a selection of recent work in urban research relevant to this theme.

Urban configurations

We noted in the sections above that distinctive spatial patterns form in urban contexts due to network effects, socio-economic distributions and cultural identifications. However, these approaches have contextualized communities in urban spaces, but have not accounted for the 'configurations' of these urban spaces. Urban configuration – a paradigm that underpins space syntax – is reflected in social uses of space as part of a relational, conceptual and physical complex (cf. Hillier, 2007). Configurations are captured through topological measurements of the urban network such as choice (relating to 'betweenness centrality') and integration (relating to 'closeness centrality') (Hillier and Iida, 2005; Hillier et al., 2010). Centralities in this way relate to distributions of land uses that influence accessibility to infrastructural networks, services and social resources. As such they have an effective socio-spatial function in urban community contexts.

However, network centralities may be settings for impoverishment and deprivation, especially in areas that are socially or economically homogenized. For example, Vaughan and Geddes (2009) analyzed the famous Booth maps of urban poverty in the 1890s to demonstrate how local integrations of impoverished neighbourhoods relate to the broader urban landscape. The spatial analytic graph shows street segments based on the intersections of axial lines, relating also to the angle of incidence of the street junction. Their analysis of London's Soho and Whitechapel neighbourhoods suggests a pattern of economically dominant streets that affected movement in the neighbourhood, resulting in low connectivity and centrality. This brought about a pattern of segregated streets within the spatial interstices, which contrasted markedly with their immediate, more affluent surroundings.

This kind of pattern, featuring pockets of urban deprivation, may reflect a paradox of opportunity in impoverished areas. Centralities may be crucial for recent migrants into slums for securing opportunities. Yet these centralities may also contribute to homogenous 'minority clustering', often based on complex economic, social and family connections within neighbourhood contexts (Vaughan and Arbaci, 2011). Hence spatial effects are inter-dependent with social and cultural factors.

Beyond the space syntax literature, significant relationships between spatial networks and other urban-context factors have been demonstrated using geo-computation, which have shed new light on the effects of distance, access and exposure in socio-spatial configurations. For example, as proximity to employment opportunities demonstrably affects employability, so travel-distances to work have a decay function, whereby less weight is given to employment in relation to distance required to travel (cf. Logan, 2012).

Significant work in geo-statistical analysis of urban networks has been undertaken by Rick Grannis (1998, 2005, 2009). Grannis focused his research upon tertiary street layouts, which are sets of streets where houses face each other, have no through traffic between them and are bounded by through-traffic streets. Tertiary (or pedestrian) streets support so-called '*T-communities*' based on frequent social interactions without the impediments of through traffic and busy roads. T-communities also tend to form chains between through-traffic streets, and Grannis has shown how homophilous racial distributions in several large American cities are based upon these street-network chains. Significantly, Grannis has shown that greater racial disparity exists *between* large T-communities than *among* them (2005); revealing how community networks based upon race form chains for as long as possible within the pedestrian street network. For this reason, Grannis (1998) has also highlighted the importance of 'trivial' streets that inter-connect blocks of tertiary streets, thus providing a path for novel community links to form.

In the field of urban sociology, Anne Power has described the significance of the urban environment for social exclusion. Examining spatially diverse communities within the patchwork of urban spatial inequalities (so-called 'jigsaw cities'), Power has described how cycles of socio-economic, material and community decline have brought about an urban pattern of discrete zones of deprivation and impoverishment (cf. Power and Houghton, 2005). The economically challenging circumstances of these zones may engender socio-cultural segregation and inter-community conflicts within their neighbourhoods (ibid., p. 195). Centralized bureaucracies may also disempower community citizens from managing their own dwellings and neighbourhoods, which perpetuates the material neglect of these places. Yet among and alongside spaces of deprivation, Power has identified spaces of community connectivity and urban innovation, which are not always visible to the outside observer (ibid., pp. 158–159).

Power's findings have been based upon a longitudinal study involving interviews with 200 families living in disadvantaged neighbourhoods. Power recorded in detail the experience of the so-called 'city survivors' who depend fundamentally upon their family and community urban networks (Power, 2007, pp. 45–46). Power found how the interactions of family members within their neighbourhood's 'inner layer' 'recreate the social vitality' of cities (ibid., p. 177). In these centralized contexts, Power has also shown how certain deprivations have stemmed from conflicts of interest among community members and urban professionals. This observation is discussed below as it relates to the theme of conceptualization.

Weisburd, Groff and Yang (2012) have presented their extensive findings of crime patterns in Seattle, based on multivariate analyses of crime and social disorder patterns within geographic micro-units. The authors recognized that most crime in the city has occurred within a highly limited number of street segments, or 'hot spots', distributed across the city. These 'hot spots' are surrounded by crime-free areas, producing a pattern of 'bad streets' in 'good areas' (ibid., pp. 186–187). Many hot spots are associated with a range of social and physical disorders, or have high mixed land use, such as those areas situated between industrial and commercial complexes (ibid., pp. 127–128).

Their findings showed how crime hot spots were associated with socio-economic indicators such as high levels of welfare benefits, school truancy, physical disorder (such as illegal dumping and substandard housing) and racial heterogeneity (which has been associated with weak community engagement). Furthermore, hot spots are associated with 'attractors' for crime, including low guardianship (ibid., pp. 110–112), unsupervised teenagers and low voter registrations, which indicate weak social controls and low intervention in social disorders (ibid., pp. 137–143). The authors also highlight the high significance of street network types in the incidence of crime. For example, many of the city's crimes occur around infrastructural assets, such as bus stops, situated on arterial roads, which inter-connect the far higher proportion of residential streets (ibid., pp. 105–110).

Considered together, the studies outlined in this section describe non-discursive agencies in the urban configurations. These relate to street networks and intersections, distance functions, homophilous aggregations and attractors for disorders such as adjacent zones within economically disparate areas. However, while each study has outlined structural or socio-cultural factors in community formations, none has considered what we term 'conceptual artefacts'. Hence, the following section draws on another sample of work in urban research that attends variously to the processes of conceptualization in urban contexts.

Urban conceptualizations

The configurations of community spaces may relate to socio-cultural identities such as ethnicity, religion, labour divisions or social class (Marcuse, 2002, pp. 11–34). Considering the separations of urban communities, the geographer David Sibley (1995) has pointed to spatially defined distinctions in the urban landscape. These, he maintains, are revealed in stereotypes of 'others' and the separate places in which 'they' live. Sibley has argued that stereotypes may provide community members with a means of coping with the instabilities of urban landscapes (ibid., 1995, p. 15). Sibley's notion of stereotypes is, we argue, one example of community members' conceptualizations of others and their significance for their local community identifications. In this way, people's descriptors of otherness in the urban environment, such as areas that are 'poor', 'rough' or 'transient' and so on, play a part in separating sets of people into discrete community areas; albeit these separations might be based upon preconceptions or even prejudices.

Community spaces that are conceptualized and defined as neighbourhoods afford the benefits of family life, social experiences and economic opportunities (Kearns and Parkinson, 2001). Yet neighbourhood identities are not spatially or temporally fixed. It is worth reiterating here how community members' concepts of their neighbourhood vary according to, for example, their age, gender, level of ability, socio-economic standing or stage in life (Lupton, 2003). Furthermore, neighbourhoods may also bring negative consequences for their community members, as in the example of young people from a particular area becoming drawn to crime; albeit the causal relationships between neighbourhood urban spaces and susceptibilities to social disorders are not well understood (Ellen and Turner, 1997).

Neighbourhoods also change their characters in different places and times. For example, the main streets and public spaces of informal settlements ('slums', 'favelas') have economic and social attractors forming community focal points, yet these might become unsafe in specific periods of the week and times of day due to traffic and conflicts among gangs (cf. Perlman, 2010, pp. 38–39).

Neighbourhoods may also be disrupted by differences in definition among community members and professionals. For example, Power and Houghton (2005) have described how policy-based interventions have served to rehouse community members in ostensibly 'better areas', leaving them with the disadvantage of separation from social cores such as family homes (ibid., p. 55). We highlight again how major infrastructural projects can undermine or replace 'community anchors', such as places for informal social encounter or for children's play (Power, 2007, pp. 58–59). Furthermore, neighbourhood relationships can be undermined by rapid social changes, not least where long-standing communities encounter 'incomers' without opportunities for inter-cultural brokering (ibid., pp. 151–152).

Considering the significance of conceptualizations in urban community formations, we may consider some methodological limitations in the notion of 'neighbourhood effects', which have focused exclusively on socio-spatial patterns (cf. Sampson, Morenoff and Gannon-Rowley et al., 2002). Lupton (2003) has argued that analysis in this regard results from a persistent separation among relevant research fields, which tend to be orientated to either individual-level surveys, or to area-level modelling of possible correlations. In order to overcome these limitations, research into neighbourhoods must accurately reflect their 'complex conceptualization', focusing on the many ways in which people and places interact, as well as the inter-relations of particular neighbourhoods within the wider urban landscape (ibid., p. 4).

Such complex conceptualizations of local community spaces have been revealed in Lynsey Hanley's authored description of growing up on a large, peripheral council estate. There she experienced a persistent phenomenon of the so-called 'wall in the head' (Hanley, 2017, pp. 148–149). This coinage refers to a mental barrier to access and opportunity, based upon conceptually internalized experiences of growing up within a spatially separated urban enclave. Moreover, this separateness includes a place-specific set of values and norms, security and self-worth (cf. McKenzie, 2015, pp. 206–207). We observe how salient aspects of the urban community space become internalized in the individual community member, transformed into values, norms and behaviours.

Conceptualizations of community spaces are formed out of actors' place-specific practices, identities and values. As such they strongly influence spatial and social behaviours. Hence developing a practicable model of community members' definitions of their localities would help urban practitioners to understand the interplay of local topo-geometries, such as centralities, with the conditions of community life, relating to 'anchors' or 'cores'. The reader might have noted that none of the studies sampled in this chapter have reflected directly upon the processes of conceptualization and definition within areas. We argue that there is a need for greater reflexivity in professional urban practice, specifically in relation to modes of discourse in conceptualization. This theme is developed in the following section.

Urban practices

The intermediaries of community spaces, their topo-geometric and topographic properties, provide the means by which urban actors both *think of* and *think with* their environments (Hillier, 2007, pp. 27–30). For example, space syntax has shown how cities are arranged topo-geometrically into foregrounds of economic movement and backgrounds of controlled, residential zones. In these contexts the observer sees the 'other' city (whether background or foreground) relative to his or her situation (Hillier and Vaughan, 2007). We *think of* these networks in terms of theoretical and professional discourse; we *think with* them in terms of quotidian actions based upon spatially embedded meanings.

We maintain a position here that *thinking of* and *thinking with* urban spaces each requires dialogue-based interactions between actors and their environments. *Thinking of* involves contrasting definitions of urban spaces, which are exchanged discursively among actors, and which compel iterative refinements and calibrations in practitioners' designs and models. *Thinking with* urban spaces are communicated non-discursively through sub-conscious exercises, yet compels dialogic exchanges of spatially embedded meanings among actors.

This dual (discursive/non-discursive) nature of 'dialogue' is reflected in principle in the work of Jack Mezirow (2000), a sociologist of education. Mezirow observed how dialogue among adult learners deals with contingencies in 'consensus-building' towards an embedded *community of practice*. Here community building necessitates freedom from coercion and distortion, so as to contextualize and weigh arguments objectively, and to regard diverse disciplinary perspectives on their own terms. Mezirow has shown how learning as an adult may achieve transformation through challenging and reconstructing the given or dominant frame of reference. Hence dialogue is a means to 'try on' the other's perspective (ibid., p. 21), to experiment using imagination and to reflect critically upon the assumptions one brings to the learning environment.

Another 'dialogic' approach to urban developments may be drawn from the notion of *urban controversies* (cf. Yaneva, 2012), which result from conflicts in description and meaning among professional and citizen practitioners. Controversies can lead to misapprehended definitions of urban phenomena, or their misapplied meanings that (as we have seen) can lead to negative effects within urban developments (Power, 2007, pp. 45–46). Attending to the need for reflexivity in practice, actor-network sociologists Yaneva and Latour have demonstrated a method for mapping professional interactions, which dynamically shape an architectural project (Latour and Yaneva, 2008, p. 87). Their approach views urban forms as generative constituencies of material and practice-based configurations, for example, the architect's drawing of a proposed development steers the thinking and discussion of clients and design professionals. Their dialogues in turn affect the architect's iterations.

Albena Yaneva has extended a social method in urban analysis, arguing that a goal of architectural theory is to achieve an 'understanding of the building as a plethora of material and subjective considerations' (Yaneva, 2012, p. 80). Yaneva argues against a set of architectural theories that uphold a 'regime of causation' (ibid., p. 33). These supposedly seek to explain those historical and cultural meanings that are reflected in, yet lie outside of, urban forms. We noted from the outset of our present discussion how practitioners avoid the reduction of society to space. Yaneva justifiably argues that space (architecture) cannot be reduced to sets of meanings, symbols and myths. Instead she argues for a pragmatic and non-reductive approach to the emergence of spatial forms from actor-network interactions. As such, Yaneva argues against any regard for descriptive correspondences of social meanings and spatial forms.

Yaneva has critiqued architectural theories that have borrowed from notions of 'social reality' in the social sciences (ibid., p. 42). Instead, she has focused upon 'oppositional' architectural theories that relate space to 'society/architecture, nature/culture, reality/rationality' (ibid., p. 39), or how we *think of* urban configurations in terms of applying theory discursively to physical artefacts. It is also possible to *think with* these configuration, to elaborate spaces non-discursively into 'socially workable patterns [. . .] through which cultural or aesthetic identities are expressed' (Hillier, 2007, p. 16).

Discussion: towards a synthesis

In this chapter we have attempted to outline some ways in which design thinking might reflect social-spatial complexities. We outlined some ways in which research in space syntax, actor-network theory in architecture and urban sociologies of deprivation have defined spatial and social inter-dependencies, addressing these to matters of urban configurations, conceptualizations and practices. We noted how community formations, and factors that might hinder these processes, often relate to physical artefacts such as pedestrian street patterns. We also outlined an actor-network theory of urban developments, which describes the social construction of the built environment.

Our sampled review has reflected upon the inter-relationships of spatial and social networks in urban communities, and how these depend upon specific conditions and externalities. We observed that spatial centralities and social cores may not locally converge (cf. Hillier, 1999; Power and Houghton, 2005). However, where spatial centralities do converge with social cores, as in the example of socially effective 'weak ties' in T-communities (Grannis, 2009, pp. 25–26), the interplay of social and spatial relationships appears to be dependent on other factors within the environment, such as cultural or racial homophily. Moreover urban community relationships may have paradoxical effects, such as providing opportunities for incomers and limiting their scope for mobility beyond their local areas (Vaughan and Arbaci, 2011).

We noted how homogeneity and separateness in the urban environment can enforce stereotypes of 'self' and 'other' (Sibley, 1995), which is perhaps a factor in the stabilization of mental models. These models may be reflected in socio-spatial structures and behaviours, including neighbourhood boundaries and place-bound identities or values (McKenzie, 2015). We maintain that community-relational dependencies also include 'conceptual artefacts', revealed in tacit definitions and sub-conscious exercises, which stem from actors' mental models. However, the nature of the relationship between a community's conceptual artefacts and a community member's set of mental models is far from clear. This relationship stems from the notion of a 'dialogic' synthesis of relational artefacts.

Considering the possibility of this dialogic synthesis, an outstanding challenge for further research in this field is to integrate configurational urban models by systematically capturing non-discursive conceptual artefacts from urban community contexts. We anticipate that this systematic capture may be achieved by extending socio-spatial analysis of an urban community to incorporate the semantic network of its inter-relationships.

Conclusion

Urban communities produce conceptual artefacts based on their members' mental models, which stem from internalizations of localized norms and values. Professional practitioners apply concepts to the urban community setting, revealed in theoretical discourses that, in general contrast to community-membership concepts, tend to be portable to other contexts. The community-membership concepts may not be revealed by the standard professional analysis and this could be to the detriment of community developments. In the following chapters, we outline some experimental methods to draw our spatial descriptions of these concepts, based on some novel applications of network modelling in urban community contexts.

Note

1 Including issues such as drug and alcohol abuse, poor mental or physical health or teenage pregnancy.

3 A case-based approach to design research

This chapter will outline methods for observing and analyzing social and economic distributions in space, in relation to urban street networks. Materials will look at some typical examples of spatial patterning relating to socio-economic inequalities (such as concentrations of deprivation), and how space syntax can be used to help understand their spatial dimensions. For example, we can often observe how high integration in the road network can 'lock in' movements in a specific section of the network. We will also look at ways in which natural and fabricated forms impact on road networks and spatial patterning, for example, the location of street networks close to riverside or peripheral industrial complexes forms segregated enclaves in the urban fabric.

Urban communities take shape in specific spatial contexts, involving complicated interplays of relationships among people and things in the urban environment (Gans, 2002, 2006; Logan, 2012). On an everyday level, people shape their urban environments as they move within 'things' such as roads, open spaces and landmarks. The movement infrastructures that connect people and things include roads, street networks, pathways and junctions across local, urban and regional scales (Batty, 2013; Urry, 2002; Sampson, Morenoff and Gannon-Rowley, 2002; Hillier and Vaughan, 2007). These scales of movement overlap along street segments, which we see when city-wide traffic convergences on local spaces. In some contexts this overlapping helps people connect to the city network, and in other contexts it gets in the way of community life. For this reason, the social and structural assets from which urban communities are formed are inter-dependent with situated, contextual urban forms (Debertin and Goetz, 2013; Gwyther, 2005; Hillier and Hanson, 1984).

Individuals and social groups often use the intelligible properties of, for example, a junction, public park, local landmark and so on, to navigate their way or achieve their sense of place. For this reason, movement infrastructures are relational to communities' needs within the urban environment, forming *boundaries*, *thresholds* and *interfaces*, with the potential to *divide*, *connect* and *allow interaction* (Palaiologou and Vaughan, 2012). Given the relational complexity of these infrastructures, the specific ways in which they impact on community life (both positively and negatively) are often overlooked by urban professionals. For example, ostensibly 'trivial' streets actually serve as significant interconnections among community groups (Grannis, 1998, 2009), and occasional places for stopping, chatting or playing become important anchors for community life (Power, 2007, pp. 58–59).

Movement infrastructures can also 'severe' communities through a 'chain of effects' relating to physical barriers and cognitive stress (Anciaes et al., 2016), bearing both conceptual and physical properties (Hillier, 2007, pp. 67–68; O'Brien and Psarra, 2015). These can include perceived boundaries that enforce stereotypes of 'us' and 'them' in the urban landscape (Sibley, 1995). Even within homogenous urban contexts, this sense of place is not fixed, but varies according to, as we have seen, a person's age, gender, level of ability, socio-economic standing or stage in life (Lupton, 2003). For example, intensively normative, class- and place-bound community formations have been observed among adolescents (McKenzie, 2015; Hanley, 2017).

The research at University College London, which formed the basis of the case studies presented in this book, sought to address some methodological challenges in understanding the 'relational complexity' of urban community spaces. The research focused on the experiences of community formations in urban spatial contexts among groups of 11- to 19-year-olds. We sampled groups across contrasting socio-economic contexts. We focused on these groups primarily because their views are often not included in formal urban planning processes (cf. UNESCO, 2002). However, their experience of urban spaces is likely to combine notions of 'rational choices' (for example, their friendships) with those of imposed rules (for example, their obligation to attend school). As such, this group would not fit some standard socio-economic methodologies for understanding urban community formations (Miller, 1992). For this reason, our research focused on the methodological challenges of observing urban community formations among young people, combing morphological, demographic and qualitative data.

The research included the configuration of some innovative visualization methods from which to observe community formations and their urban contexts; they are outlined below. The participatory section of the research revealed that movement infrastructures feature as one type of artefact among many others within the sampled community formations, across all age groups and socio-economic contexts. Other dominant features included open spaces, such as public parks and leisure facilities, shopping areas, industry and sites of pollution, and local schools. We focus the present analysis and discussion on what these methods might show about impact of movement infrastructures (roads, streets, junctions and pathways).

Working with space syntax

In order to model the possible impacts of movement infrastructures, the investigators generated a series of morphological models using space syntax. Space syntax is, to reiterate, a theory and method that accounts for spatial configurations in relation to patterns of socio-economic activity and cultural meaning (Hillier and Hanson, 1984; Hillier, 2007). Its methods allow researchers to test conjectures about urban spatial movements, relationships and meanings based upon 'non-intuitive' actions such as reasoning, induction and analysis (Karimi, 2012). Here we focus on a selection of theories from space syntax literature that lie at the heart of our synthesis: the notion of relational complexes, the structures of network centralities and of foreground and background networks and the possibility of non-discursive methods.

Hillier has described how the built environment comprises 'relational complexes' that constitute its buildings and cities (Hillier, 2007, p. 74), built out of object and abstract artefacts (ibid., p. 67–68). As object artefacts, such as streets and buildings, are subject to natural and physical laws, so abstract artefacts include the realizations of 'socially meaningful configurational entities' that are subject to spatial functions or significant rules (ibid., p. 74). For example, a neighbourhood boundary may be shaped around a set of streets (object artefact), where the rules for what is inside and outside that boundary are based on the local population's sub-conscious or tacit delimitation of its community space (abstract artefact).

Space syntax has not to date considered how abstract artefacts also relate to conceptualizations of space by community members. Hence we introduce a novel term, 'conceptual artefacts', to encapsulate the spatial products of non-discursive agencies or sub-conscious exercises in urban use patterns. These may be revealed in sets of tacit assumptions and meanings that are subsumed in quotidian activities, influenced by community members' 'mental models' of that space. Analysis of mental models that underpin the formations of these artefacts would require an empirical method to inter-connect urban configurations with spatially and socially embedded meanings.

Space syntax has a core technology in the computational modelling application, Depthmap. The axial map of the urban network is presented to this application (by way of an imported drawing), and the software is programmed to recalculate the map based on two network-analysis measures of closeness (e.g. how far one element is from all others) and betweenness (e.g. how connected one element is to all others). Space syntax has adopted a novel terminology for these measures, which are 'integration' and 'choice', respectively. Hence, integration is the measurement of origin-to-destination distances, representing the likelihood that any structures underpin movements *from* or *to* a location in the network. Choice is the measurement of flows along those certain structures, representing the likelihood that they underpin movements *through* the network. Distances among network segments are also weighted by the angles between segment intersections (the shallower the angle for turning from one street segment to another, the higher the natural movement). This weighting stands as a proxy for 'natural movement' across the road network, which we review in greater depth in Chapter 4.

The Depthmap application provides a visual model of any sampled urban street network (or architectural space), based typically on a warm-to-cool or heavy-to-light gradient ramp to represent high-to-low values for movement potentials across the network. Depthmap handles two bespoke calculations: 'choice' represents the probable affordances for movements 'through' the network; 'integration' represents affordances for 'origin/destination' movements. A useful example of space syntax in practice is in the representation of a city's integration core, which shows its overall 'centre of gravity' in terms of movement potentials. We provide case-study examples of these in Figure 3.3. Low-to-high values are shown on a light-to-heavy line gradient ramp, which represent the quintile range in which the integration value of street segment falls.

Integration and choice are calculated based on the distances among connecting paths that radiate from the iteratively sampled segment. Depthmap processes are typically based on a set of radial scales for the computation of integration and choice. For example, 500m representing a local scale, 10km representing a city-region scale, or else the global network scale. Readers may see the radial scale annotated such as, for example, R500 (local scale), or Rn (global scale). Moreover, integration and choice may be normalised to allow comparison among different model samples, and these are often annotated such as, for example, NAIN_R500 (normalised integration at a radial scale of 500m), or similar.

The Depthmap application is available to freely download and use[1] and will work on most up-to-date personal computers. Users can import road-network data sets into Depthmap, which can be sourced either from a geographic data repository such as Geofabric.de (for Open Street Map data), or else can be manually traced over a digital map or blueprint using a CAD or GIS application. Users can then export a digital drawing from CAD or GIS (for example in .dxf or .mif format), ensuring that each of the elements in the drawing includes a unique identifier such as a row number. Any problems with importing a drawing to Depthmap might be to do with missing element identifiers. Once the drawing has been imported to Depthmap users should save it as a graph file, which will help to prevent application data loss during processing. Depthmap processes the digital drawing data in two stages. Firstly, the axial map calculates pathways based on concatenations of elements. Secondly, the segment map process calculates sections within the axial lines based on their network intersections with other lines.

Community research into spatial patterning

The aims of the research, upon which the case-studies were based, posed a methodological challenge in bringing together an understanding of community perceptions of their local spaces and the spatial dynamics of the wider urban network. The researchers addressed this challenge by gathering community perspectives through participatory workshops (described in detail in O'Brien et al., 2016), and by using space syntax as a theory and method of urban morphological configuration, based on the notion of affordances for 'natural' movements across street networks. These affordances are termed 'movement potentials' within the space syntax literature (Hillier and Hanson, 1984; Hillier and Iida, 2005). The scripts generated to display the findings of the workshops and applied network models, as described throughout this chapter, can be freely accessed via an online repository.[2]

Area of study

Our aim in this section of our research was to compare community formations in contrasting contexts. We selected field sites in Liverpool, UK, which offered opportunities to address comparative experiences of community formations in transforming urban environments. The sites' different geographic and social characteristics provide contrasting contexts to our study (Robinson, 2016).

The maritime city of Liverpool is a non-capital, medium-sized conurbation, presenting morphological characteristics of grid-pattern, inner suburban street networks and low-density peripheries. The city bears radial road networks that have formed in relation to riverside industries, and these have shaped historically its spatial layout (Serra et al., 2018; O'Brien and Griffiths, 2017). While enjoying higher-than-average economic growth in the period 2009–2014 (Liverpool City Council, 2016), the city has experienced among the highest levels of multiple deprivations of any UK local authority (Liverpool City Council, 2015). The Liverpool region is 'a place of contrast and social and spatial disparities' (Sykes et al., 2013), bearing a range of spatial inequalities reflected in zonal concentrations of wealth and poverty (ibid., p. 6). The city region can be characterized historically as an area of prolonged industrial decline, which has been reflected in higher-than-average unemployment, neighbourhood dereliction (Leeming, 2013) and low business density (Liverpool City Council, 2013). There are reportedly signs of economic renewal in the region (Liverpool City Council. 2016), with its economy and innovation developing at a higher rate than the national average.[3]

To observe aspects of urban community formations within these urban contexts, we selected areas to represent a range of socio-economic contexts. Figure 3.1

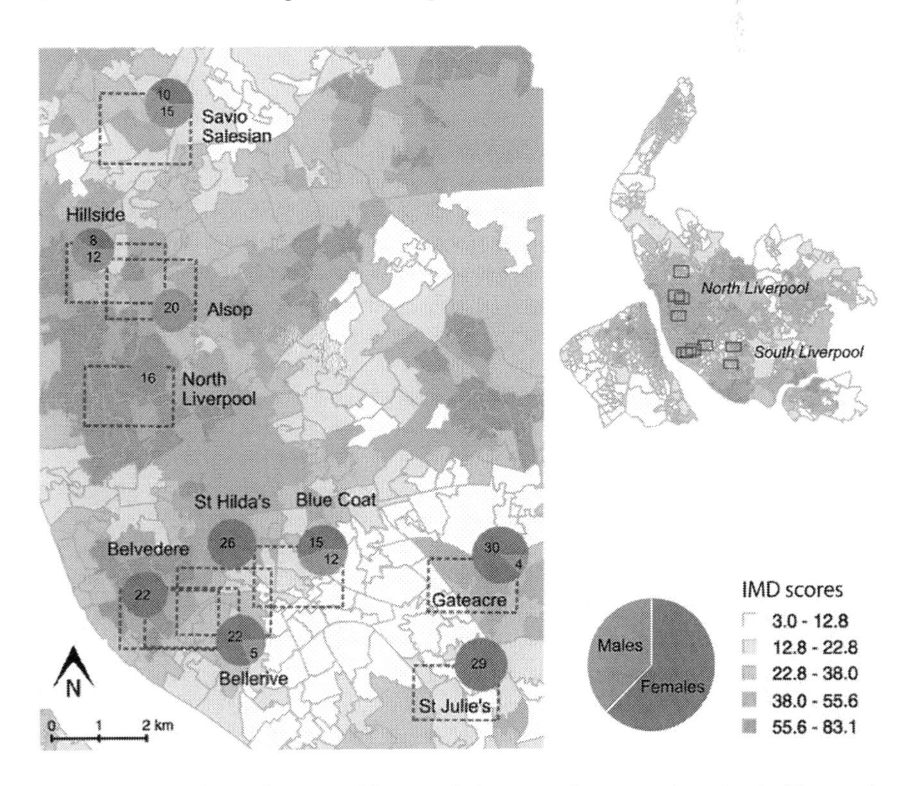

Figure 3.1 Locations of communities sampled over socio-economic status in Liverpool, UK (based on IMD; low-to-high values shown on a dark-to-light ramp). Pie graphs displaying number and distribution of participants in each workshop site.

shows the areas where the map-making workshops were conducted in Liverpool. These workshops were undertaken at schools within zones presenting a range of Index of Multiple Deprivation (IMD) scores (DCLG, 2015). These statistics are organized by Lower Super Output Areas (LSOAs) that represent surveys of approximately 800–3000 people (a detailed description of the selection is provided in O'Brien et al., 2016). These provide measures of relative deprivation in the surroundings of each school. Note, for example, in the figure, how the North Liverpool Academy community coincides with moderate to high multiple deprivations, while the Blue Coat community coincides with moderate to low multiple deprivations.

In total, 246 participants engaged in the workshops, with 34% male and 66% female students. The amount and gender distribution of the surveyed students are also depicted in the pie charts of Figure 3.1. Note a bias in the sampling towards female participants, which was the result of voluntary engagement with the project and not of deliberate sampling.

Community workshops

In gathering data pertaining to the community perspective, we adapted the ethnographically based approach of community appraisal, which helps to equip participants in analyzing and describing their own 'realities' (Chambers, 1997). We devised and ran a series of workshops located at school facilities in Liverpool. The workshops engaged the participants in brief introductions to urban planning and design using a city-scale map, which helped their understanding of how interventions impact on community life. As most participants were children, it was not possible to gather data about individual circumstances. However, an assumption was made that the participants had general experience of typical social, structural and environmental characteristics of their local areas.

The research participants worked with local-scale maps, in which the cartography located the workshop site at the centre of an approximately 2km^2 area. Participants used well-known 'emoticon' symbols (Figure 3.2) to associate basic emotions and experiences in specific urban environments. The participants were free to represent and locate using the provided symbols (see further details in O'Brien et al., 2016).

Figure 3.2 Emoticon stickers used in the Liverpool map-making workshops.

Following the selection of the features of the local areas that deemed significant for community formations in both case studies, all participants were invited to fill out a table to list and describe the emoticons and icons they used on their maps. For example, a 'shocked face' emoticon relating to a major junction was described as 'Can't get across, too much traffic!'. This allowed the investigators to identify structures or places in terms of their affordances and hindrances for social connectivity.

Finally, all the maps were scanned into a GIS and points data were digitalized and tabulated by gender, age group, urban type ('road', 'open space' and so on), the specific name of the selection and any text description where this was available. All data was aggregated to reveal distributions of points. For example, the researchers were able to plot maps of community assets for which emoticon stickers represented their respective positive and negative connotations. This was to guarantee consistency on the comparison of the case studies.

Results

Relevant places selected by community

We sought to understand how the features selected by the participants in terms of their relatedness in space from the community perspective. We applied K-nearest neighbour (Knn) graphs to the data, where the nodes represent the selected local features and the edges the four closest neighbours (see Figure 3.3). The graph nodes are geo-located based on the centroids for each cluster of iconographic symbols arranged by name (for example, the centroids for all points with name 'Princes Park'). The node sizes were adjusted in proportion to the count of icons applied to each feature (%). The Euclidean (shortest path) graph edges represent the 'semantic' distances among the features selected. For example, participants at Blrv and StHld have attached high significance to local open spaces (in Figure 3.4 they are 1. Sft_Pk and 2. Prn_Pk). Alsp and Hlsd have attached greater significance to nearby movement infrastructures (in the figure they are 3. County Road and 4. Breeze Hill), which in each case seem to mark a functional separation among groups of features within the community's local space.

From the Knn graphs, the researchers were able to observe how some major roads played a role in community formations. In the next stage of the research, we sought to understand how these roads are configured within the local community space, including their interactions with the city network. This involved analysis of urban morphological integration at local and city scales based on the Depthmap measurement of normalized integration (NAIN). To reiterate, the integration model demonstrates the likelihood of network segments affording origin-to-destination movements within a certain radius of network configuration. NAIN can be seen as providing a model of spatial accessibility within the network.

The researchers generated urban morphological models based on integration at radius 400m, 2000m, 5000m and global scale (which were then normalized

Figure 3.3 Urban integration cores for the Liverpool, UK, conurbation. Local scale based on 400m (left), and global scale based on 5000m (right).

North Liverpool

South Liverpool

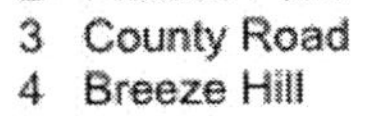

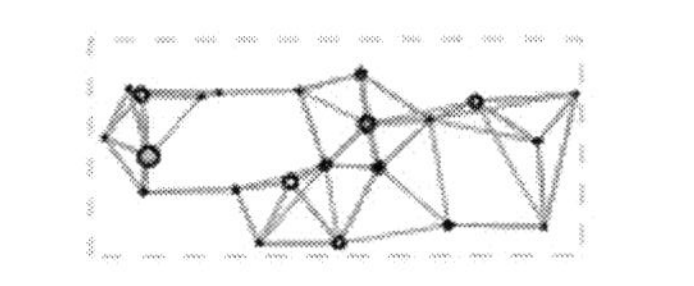

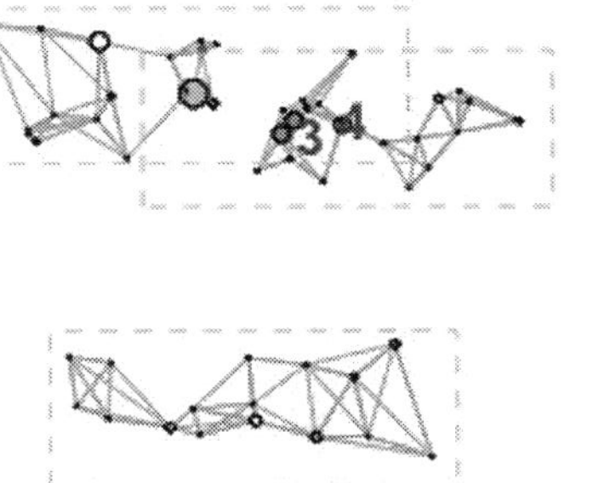

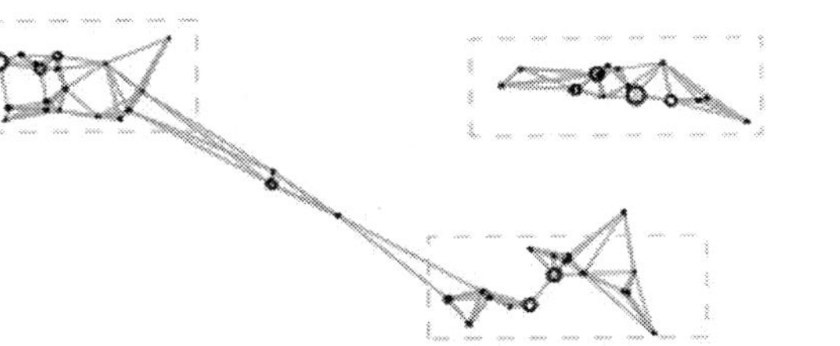

Type of Local Features

- railway
- school
- cemetery
- roundabout
- underpass
- junction
- transport
- road
- residence
- shops
- area
- cafe/restaurant
- leisure
- open space
- police station
- library
- crossing
- sports
- emergency services
- nursery & children's centre
- stadium
- alleyway
- industry
- gardens
- place worship
- woodland

Count of Icons

- 4 - 5
- 5 - 9
- 9 - 19
- 19 - 250

Figure 3.4 K-nearest neighbour graphs of the Liverpool group of sampled communities.

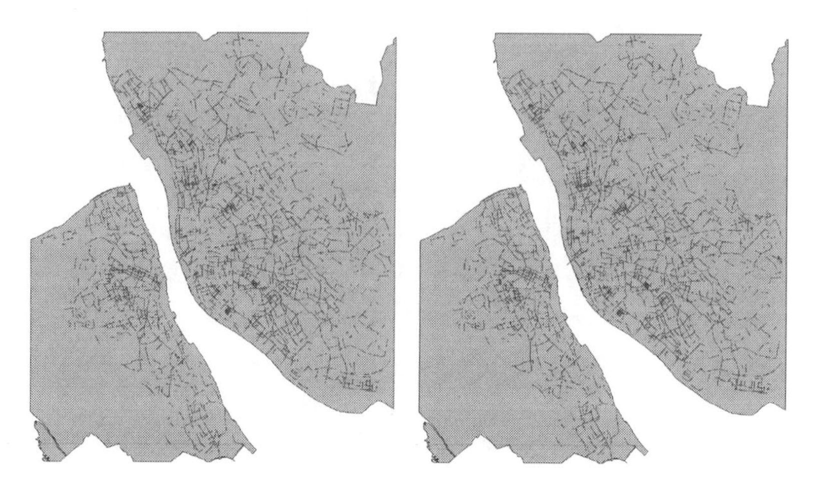

Figure 3.5 Urban morphological models for Liverpool and the Mersey conurbation sample, based on integration at scales of R400 and global radii. The maps represent the highest affordances for origin-to-destination movements at these scales (based on the top 20% integration value range).

for comparison; see the examples in Figure 3.5). We then extracted the range of network sections at which high-value local segments (400m) overlap with high-value global-scale segments (Figure 3.6). The purpose of this was to observe the 'potential' impacts of wider-scale movements on local-scale centralities.

Impacts of infrastructures on community formations

The arrays of community sample maps (Figures 6.1 and 6.2) reveal how icons with 'positive' and 'negative' connotations were applied to movement infrastructures. Communities that share network spaces (such as Blrv, Blvd and StHld, and also Hlsd and Asp in Liverpool) present very similar impressions of their local spaces, with each community applying similar weightings to major local infrastructures, such as busy roads, junctions and pathways. The weights applied to these movement infrastructures were in the main based on negative representations (of 'shock' or 'sad' and so on).

Perhaps surprisingly, the majority of icon weights were applied to movement infrastructures that do *not* tend to converge NAIN local and global scales. We might hypothesize that the communities' respective 'sense of place' (whether based on negative or positive connotations) is affected by movement potentials bearing singular, dominant NAIN-scale segments. We can look at this pattern in different ways. One possible interpretation is that movement infrastructures that converge NAIN scales repel community life. Another is that the participatory methodology has revealed how 'negative spaces' of community life form around non-convergent movement infrastructures.

The researchers' aim in this section of work was to take a broad survey of the wider urban contexts in which our sample populations were situated. We based this

Figure 3.6 Segments at which NAIN R500 segments overlap NAIN R5000, for the Mersey conurbation sample.

survey on measurements of 'potential' impacts of city-scale centralities on local centralities, in relation to spatial distributions of relative deprivations. We sought to understand how overlaps among local and global NAIN scales might relate to distributions of relative deprivations or vulnerabilities. We did this through the use of the Index of Multiple Deprivations (IMD; DCLG, 2015).

The researchers measured the total length of segments that feature overlapping movement potentials within square subdivisions (40 x 40 grid), as a percentage ratio of the total length of all segments in each square. This allowed us to measure the degree to which movement infrastructures in a local area interacted across radial distance scales. For example, certain areas might feature high-density street networks, but have few segments interacting with city-scale movements.

Figure 3.7 shows for Liverpool a sample of the results for the square subdivisions within each Lower Super Output Area (LSOA). A choropleth array was

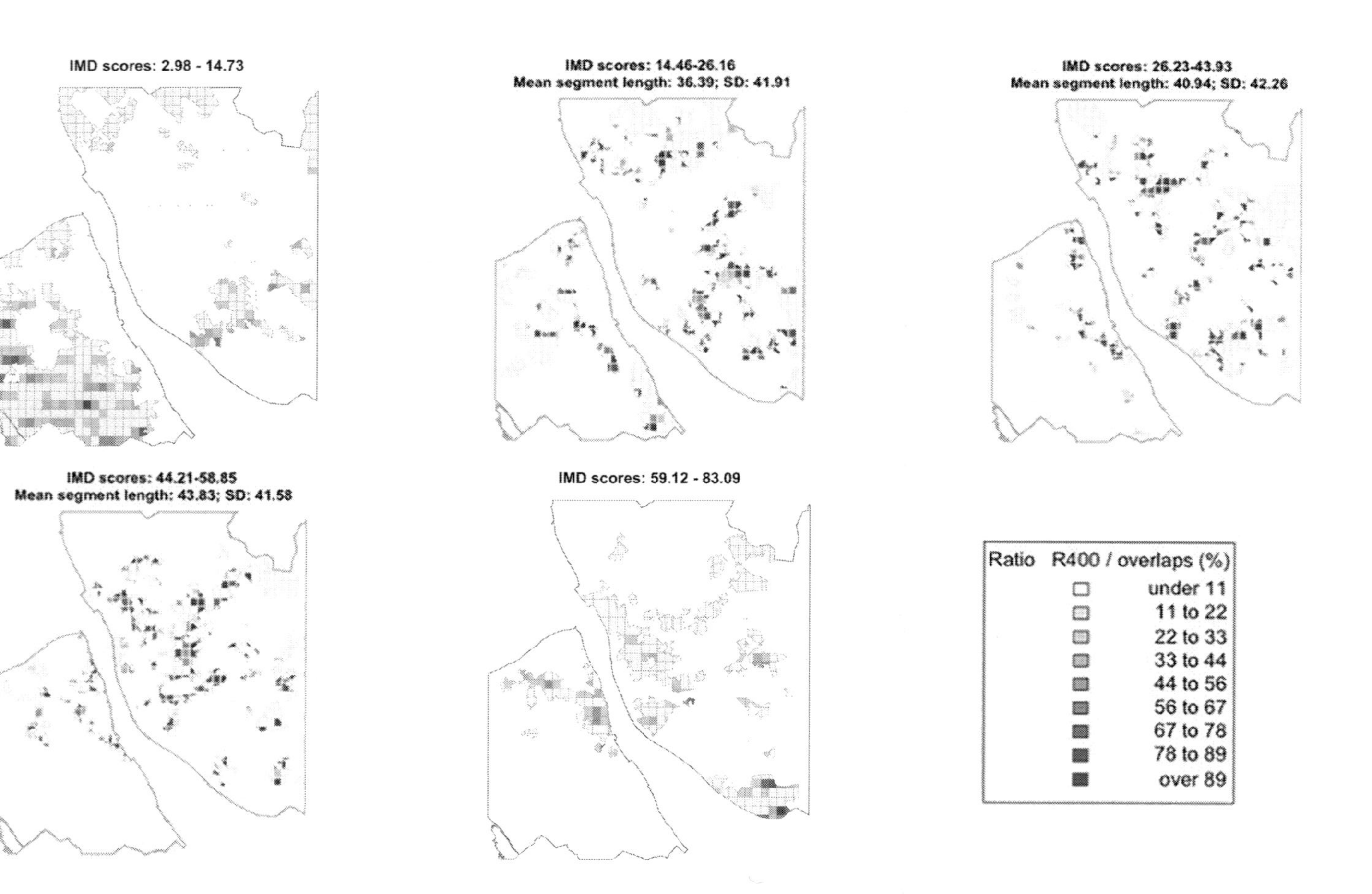

Figure 3.7 A choropleth array representing total lengths of network segments with 'overlapping' scales of potential movements.

generated to represent the segment length counts broken down by IMD score quintiles, which vary across a low-deprivation to high-deprivation range of 2.96–82. In the figure, the distribution of network interaction lengths is graduated from light (low length count) to dark (high length count).

Discussion

In this chapter we have outlined some methods for observing possible impacts of movement infrastructures in urban community formations. In Chapter 2 we argued that urban communities make use of artefacts in their local spaces that, by combining physical and relational properties, are relational to the activities and perspectives of the community members. In this chapter we sought to observe how the major presence of roads, streets and pathways in community spaces served to shape the 'sense of place' in that area. We drew from three sources of data: geo-located points data derived from participatory workshops, socio-economic data from public sources and urban morphological data pertaining to probabilistic urban-network interactions.

We also observed the patterning of convergences of integration scales in relation to distributions of relative deprivation or vulnerability. We found a degree of correlation in the two case-study contexts of Liverpool. In Liverpool, high-deprivation areas tend to converge integration scales. Discounting these indicators of socio-economic status, the convergences seem to relate more to urban morphologies based on dense, rectilinear structures. Areas bearing sparse, non-linear structures such as those typically found at the urban peripheries tend to have a singular, dominant integration scale. These perhaps represent 'islands' of communities that are segregated in terms of relative deprivation. We draw a hypothesis from this study that movement infrastructures bear highly variegated significance for community formation, depending on the communities' specific constitution and circumstances.

Notes

1 Update on versions and support is available via https://github.com/SpaceGroupUCL/depthmapX.
2 Available to download at www.routledge.com/9781138652057
3 Source: https://www.liverpoollep.org/growth-sectors/ [accessed 21st March 2019]

4 Building on the past

This chapter will outline methods for observing the development of road networks over time. This is valuable for understanding the historical depth of road networks and their significance for contemporary community formations. The chapter will develop a case study of city development from its origins, exploring ways in which historical spatial dynamics have shaped its road network. We will observe how subsequent stages of development involved densification and extension of networks across the local region. We will see how development in the industrial period brought about high-density street layouts based on ready access to economic centres, and also how social inequalities were (and are still) reflected in the geometric orientations of sub-networks (the geographic theme of 'power geometries'). We will then look at the impacts of 'modernist' projects, such as major road building initiatives, in terms of relationships between regional and local connectivities. We will see how major roads have the effect of privileging city-scale movements from suburbs to metropolitan centres.

We describe some observations of Liverpool's (UK) syntactical patterning relating to its urban network growth from the 1850s to the present day. Liverpool's rapid growth and transformation provides a compelling case study of network configurations as they relate to movement potentials over time. We argue that syntactical analysis of movement potentials provides a tool for evidencing urban historical socio-spatial patterning. Urban patterning might be shaped *variously* by historical factors such as socio-economic inequalities, labour divisions, ethnicities or religious denominations. We have attempted to demonstrate how movement potentials have persisted *normatively* along structural path-dependencies that underpin these patterns.

This study was based on samples of three prominent centralities of Princes Avenue, Scotland Road and Canning Place, across four periods: 1850s, 1890s, 1950s and contemporary. We prepared Depthmap data samples using an arrayed visualization format, which allowed us to make comparative observations of movement potentials as they converge and intersect across various urban scales. This has allowed us to generate 'internal' perspectives on configurations over time, to suggest some possible effects of city-scale morphologies on local spatial dynamics.

Introduction

Britain's rapid industrialization was driven by mass manufacturing across Lancashire coupled to Atlantic trade via the port of Liverpool (Hobsbawm, 1962, pp. 49–50). By the mid 19th century, the city's extensive wharf complex carried the world's highest volume of cargo (Aughton, 2008, p. 217). Much of Liverpool's economic activity was focused along this complex (only a minor manufacturing area was centred around the Ropewalks area). The city's high-density urban networks radiated outwards from the riverside, wherein some localized centralities supported trade and consumption. Massive inward migration from across the British Isles also led to highly segregated communities based on religion, nationality and work. Irish Catholics made up largely unskilled labour sub-groups around Scotland Road. Welsh Presbyterians made up semi-skilled labour groups around Princes Avenue. English and Scottish Protestants made up skilled labour and clerical groups also around Princes Avenue and in suburban areas (Pooley, 1977; Lawton, 1979). See Figure 4.1 for a map showing the main locations referred to in this chapter.

By 1911, when construction of the landmark Royal Liver Building was completed, the port city had reached its peak of growth and innovation, followed by a decades-long process of general decline. The Depression of the 1930s led to one-third of working-age men being out of work (Aughton, 2008, p. 244), and wartime bombardment and displacement led to intensive damage to the city's fabric. In the post-war years, road building, slum clearances and mass rehousing led to a precipitous decline in population within a severely post-industrial landscape (cf. Sykes et al., 2013).

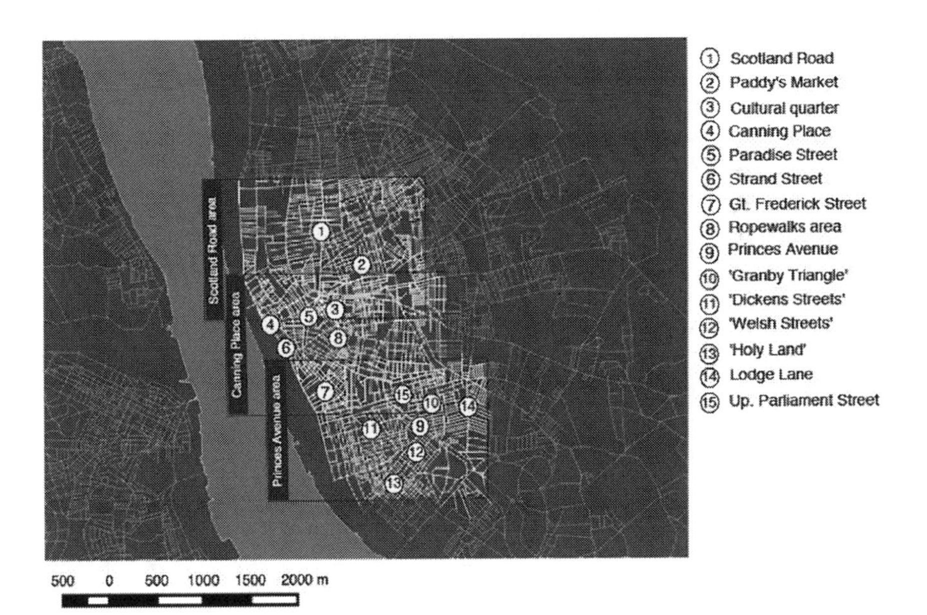

Figure 4.1 Annotated map of Liverpool's street network in the 1890s, showing areas sampled and key locations described in the study.

High unemployment and social exclusions led to the major Toxteth Riots of 1981, which furthered the distress to Liverpool's urban fabric. In spite of recent urban developments, Liverpool has experienced among the highest levels of deprivation in the UK, measured by the Index of Multiple Deprivations (IMD), featuring extensive zones presenting high unemployment, high rates of chronic illness, low rates of educational attainment and low longevity (Liverpool City Council, 2015).

The urban history of Liverpool provides a compelling case study in rapid industrial transformation and accelerated decline. Tracing the distinctively 'urban' evolution of a city poses a range of epistemological dilemmas, which we have attempted to resolve to some extent in this chapter. Firstly, do we deal with the city as an autonomous or as a complimentary entity within its wider spatio-temporal context (Jansen, 1996)? Secondly, how do we reflect the general 'problem' of urban space as, at once, the locus of everyday living and the site of normative (cultural and ideological) reproduction (Rodger and Sweet, 2008; Lefebvre, 1991, p. 50; Hillier and Hanson, 1984, p. 22)? Thirdly, how do we accommodate the paradox of the city as both a compressor of space-time, in terms of rates of flows and exchanges, and as an instrument for socio-spatial distancing, in terms of social differentiations expressed in spatial patterning (Dennis, 2008; Massey, 1994, p. 147) and relational complexity (Hillier, 2007, pp. 67–68; O'Brien and Psarra, 2015)?

Space syntax offers a method to address dualities in urban formation through its 'configurative' model of morphological generation, whereby generic spatial patterning relates dynamically to characteristic social practices (Hillier and Vaughan, 2007). This approach has been successfully applied to a range of 'industrial age' urban histories (for example, Psarra et al., 2013; Griffiths, 2009, 2017; Al-Sayed, Turner and Hanna, 2009; Pinho and Oliveira, 2009; Vaughan, 2007; Vaughan and Penn, 2006; Medeiros, de Holanda and Trigueiro, 2003). However, Hillier and Hanson's configurative theory, from which our methods of analyses have been extended, has been critiqued for its epistemological perspective of an 'external observer' of urban space (Griffiths, 2011). Instead of this theoretical viewpoint, we take a configurative perspective of the 'internal observer', whose situated space-time coordinates determine a localized (relativistic) modality for spatial description. The 'internal perspective' is pertinent to urban historical analysis when we conceive of how flow rates through urban networks in the past related to the capacities of their dominant mobilities (Urry, 2004). For example, the movement-scale paradigm of a predominantly pedestrian milieu offers affordances for spatial description that are qualitatively differentiated from those of an automotive milieu. Moreover, these local spatio-temporalities set in place basic components for the urban system's spatial evolution. As such, we cannot describe an urban history without considering the 'clocks' of its landscape (Ingold, 1993).

Urban historians are presented with a challenge in evidencing relativistic spatio-temporalities (Pooley, 2000). Colin Pooley (2016) has reviewed, for example, a diverse set of historical materials relating to transportation-in-use within British conurbations. From the evidence available, it appears that most urban inhabitants from the mid 19th century walked or took horse-drawn and, later, electric omnibuses. Suburban inhabitants in the early 20th century depended on electric omnibuses and, later, on personally driven automobiles (which from the later 20th century also introduced social inequalities of inclusion and accessibility).

Another approach to this challenge might be sourced from the space syntax notion of *natural movement potential* (Hillier and Iida, 2005), which encapsulates a method to recover *probabilistic* urban movements from the 'internal observer' perspective. Natural movement potentials (represented through space syntax measurements of choice and integration) may approximate the inter-relationships of these movement scales and network morphologies, as well as interactions between scales such as those between local through-movement and global to-movements – while recognizing that labels such as 'local' and 'global' are normative and negotiable over time. Space syntax can provide the urban historian with 'potential' evidence that engages in new ways with traditional data sources; for example, in relating street networks bearing high-integration values to census data reflecting demographic segregation in those areas – as was typical of 19th-century Liverpool (Pooley, 1977; Lawton, 1979). In this chapter, we mainly explore space syntax as a theory and a method for breathing new life into such sources, offering richer interpretative possibilities than conventional GIS visualization techniques (Griffiths, 2012, 2013; Vaughan and Penn, 2006).

Data sets and methods

In investigating Liverpool's historical urban network we sampled three major centralities converging around Princes Avenue in Toxteth to the city's south, Scotland Road in Everton to the north and Canning Place in the city centre. The urban networks were modelled by tracing Ordnance Survey Ancient Roam tiles using QGIS. The tracings were exported to Depthmap for recalculation with angular segment normalized choice and integration measures (NACH and NAIN), based on metric radii of 200m, 400m, 800m, 2000m, 5000m and 'global n' scale, respectively. We sampled from four historical periods: the 1850s marking the height of Liverpool's first wave of expansion; the 1890s as the period approaching the city's economic zenith; the 1950s marking the city's network maximum; and the contemporary network subsequent to major remodelling and developments. In addition, we generated a Depthmap model (NACH and NAIN, R100, R250, R400, Rn) to reflect urban redevelopments to the city centre in the 1970s.

This diachronic sampling constitutes one section of the broader project at University College London, 'Visualizing Community Inequalities'. In this section, we seek to describe our approach to contemporary community formations in relation to persistently 'imageable' urban forms, such as boundaries, interfaces and thresholds (Conroy-Dalton and Bafna, 2003; Lynch, 1960). To achieve this kind of observation, we generated Depthmap model arrays using an R programming environment (RStudio) to make comparative observations 'within the eyespan' (Tufte, 2006). This comparative method is reflected in our citing throughout text of the radial scales from which we made our observations (e.g. 'NAIN 400', 'NACH 5000' and so on). We provide an overview of the diachronic samples in Figures 4.2 and 4.3.

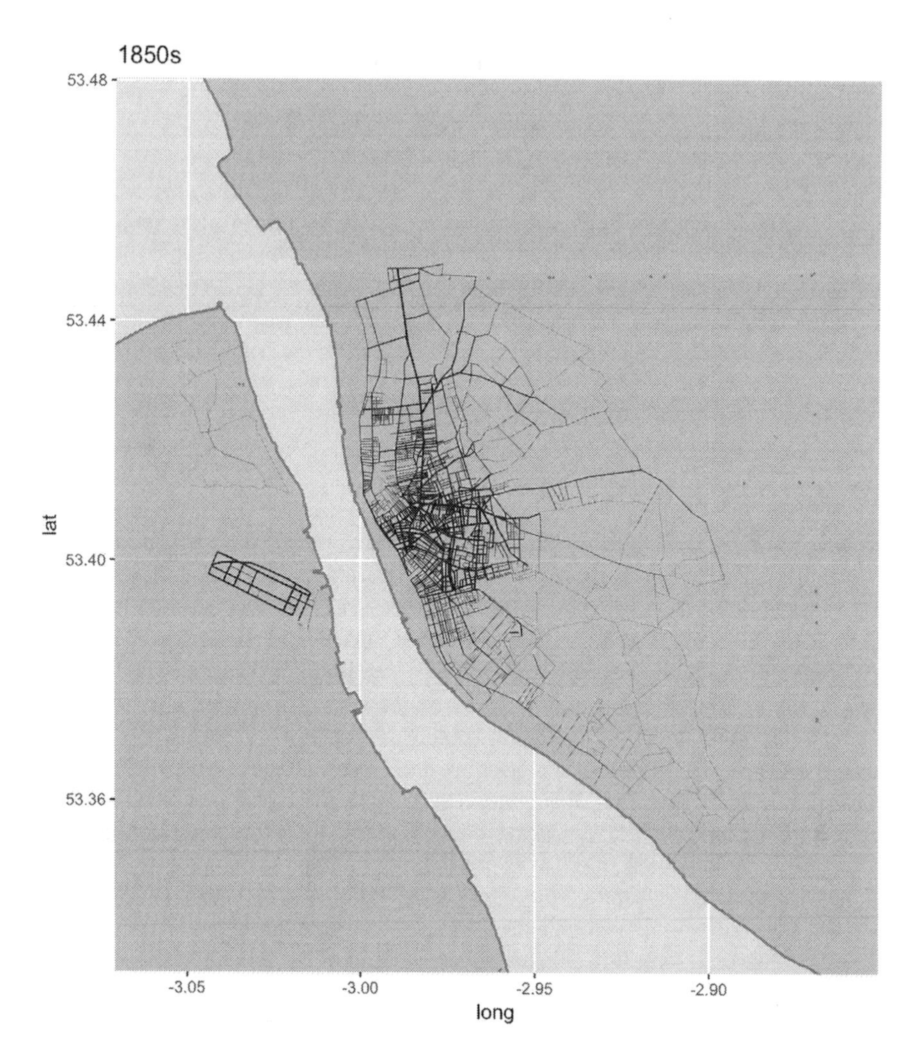

Figure 4.2 An array of Depthmap models of Liverpool, UK, representing the syntactical evolution of urban regional integration over four historical periods (NAIN Rn): 1850s, 1890s, 1950s and contemporary. Segment line widths have been graduated from high (heavy line) to low (light line).

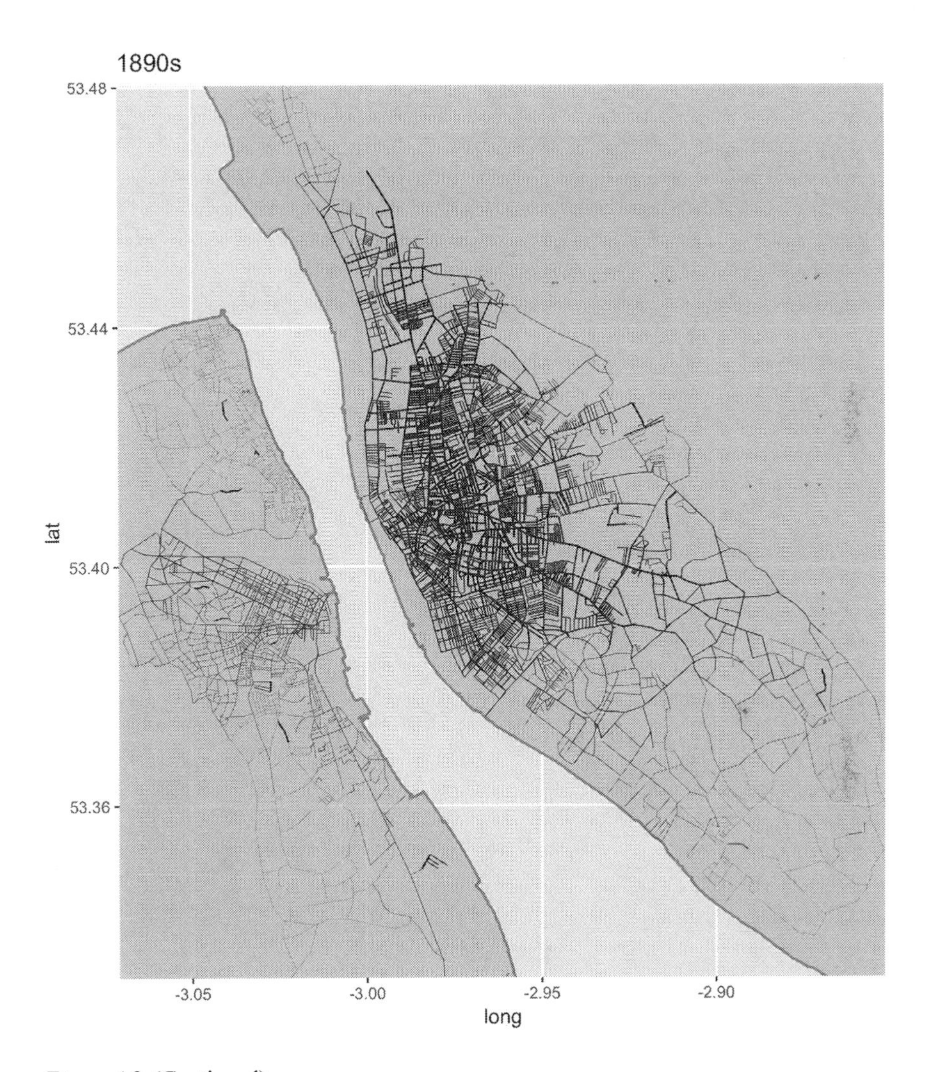

Figure 4.2 (Continued)

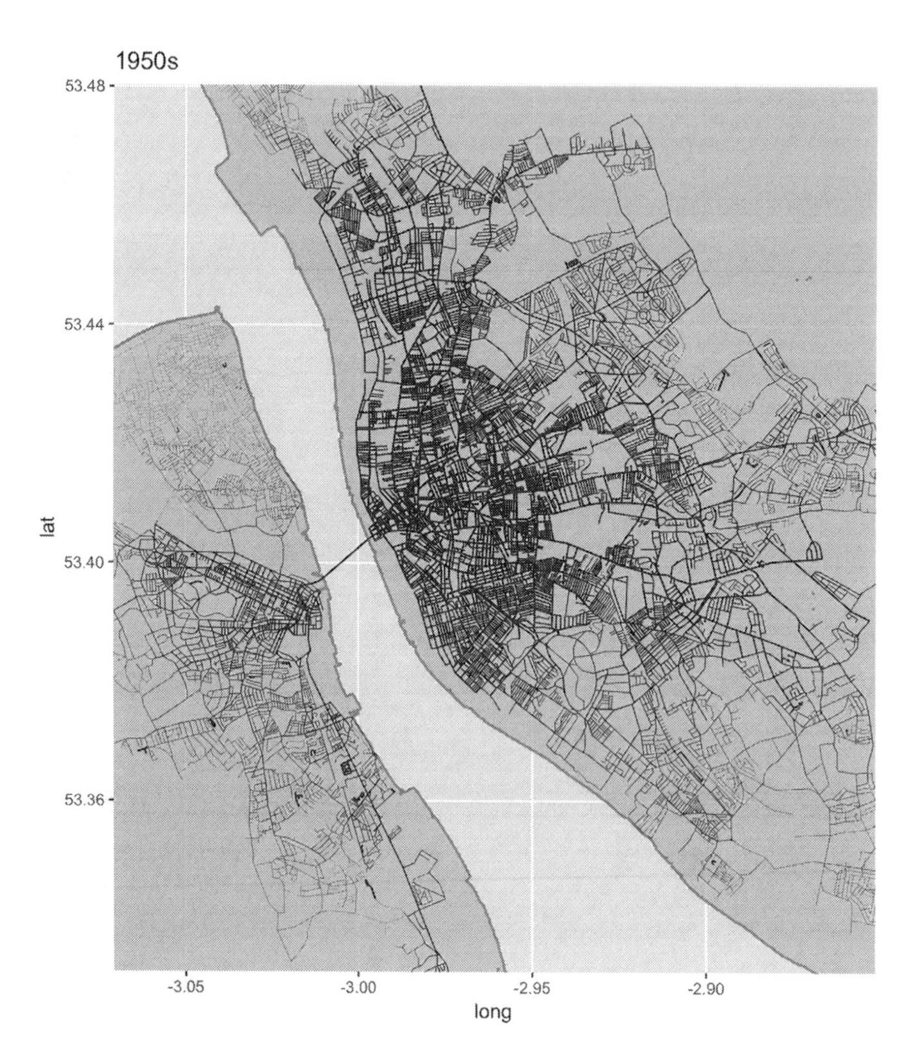

Figure 4.2 (Continued)

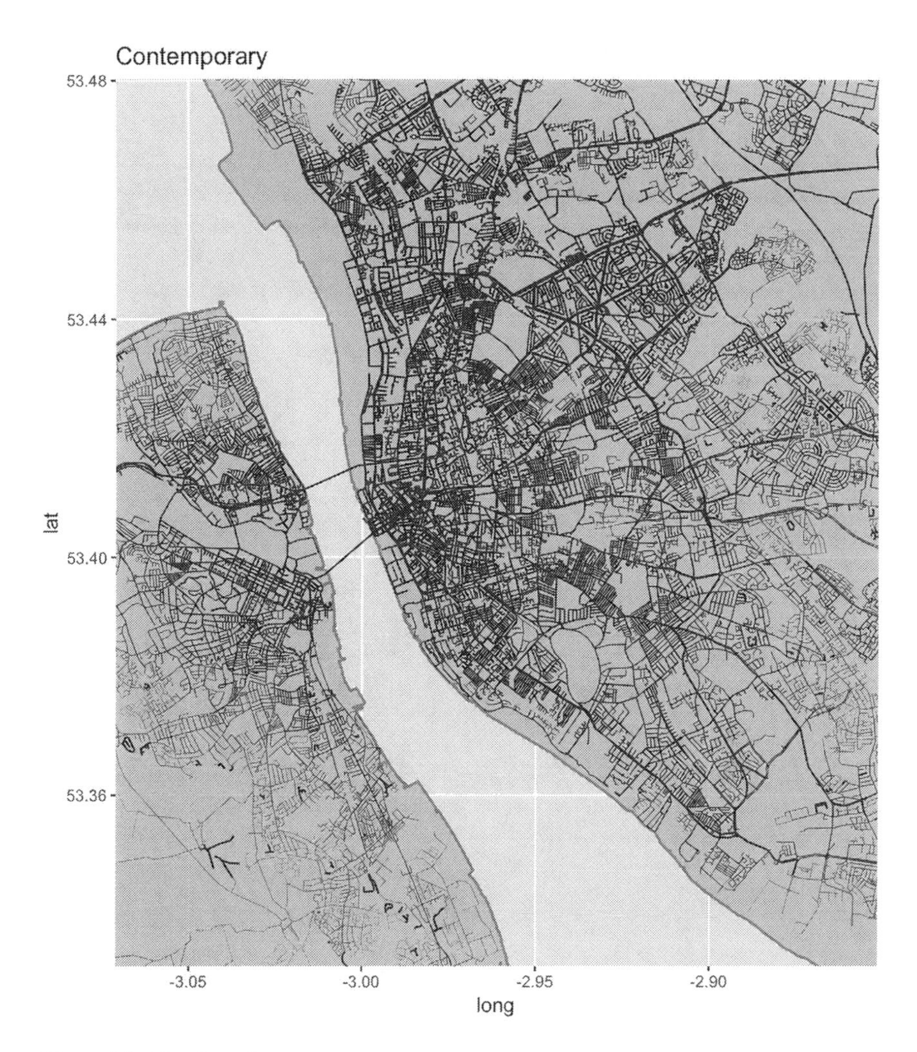

Figure 4.2 (Continued)

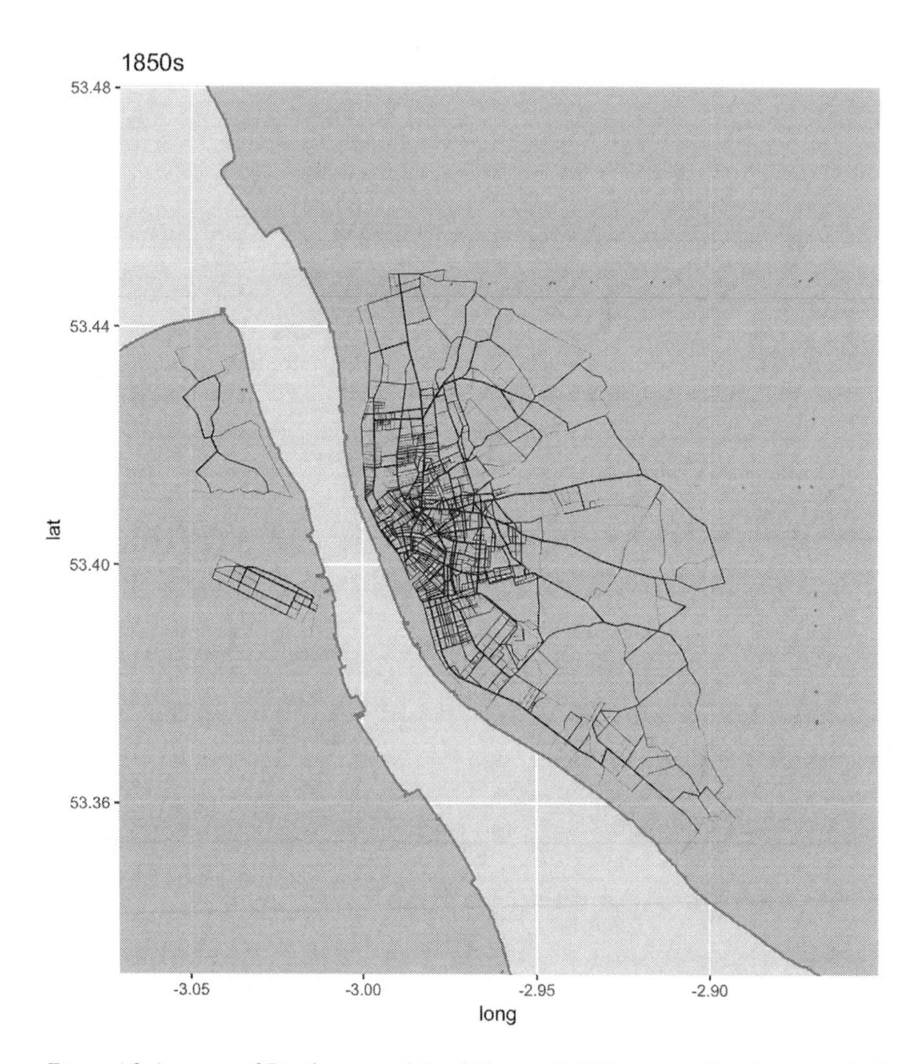

Figure 4.3 An array of Depthmap models of Liverpool, UK, representing the syntactical evolution of urban regional choice over four historical periods (NACH Rn): 1850s, 1890s, 1950s and contemporary.

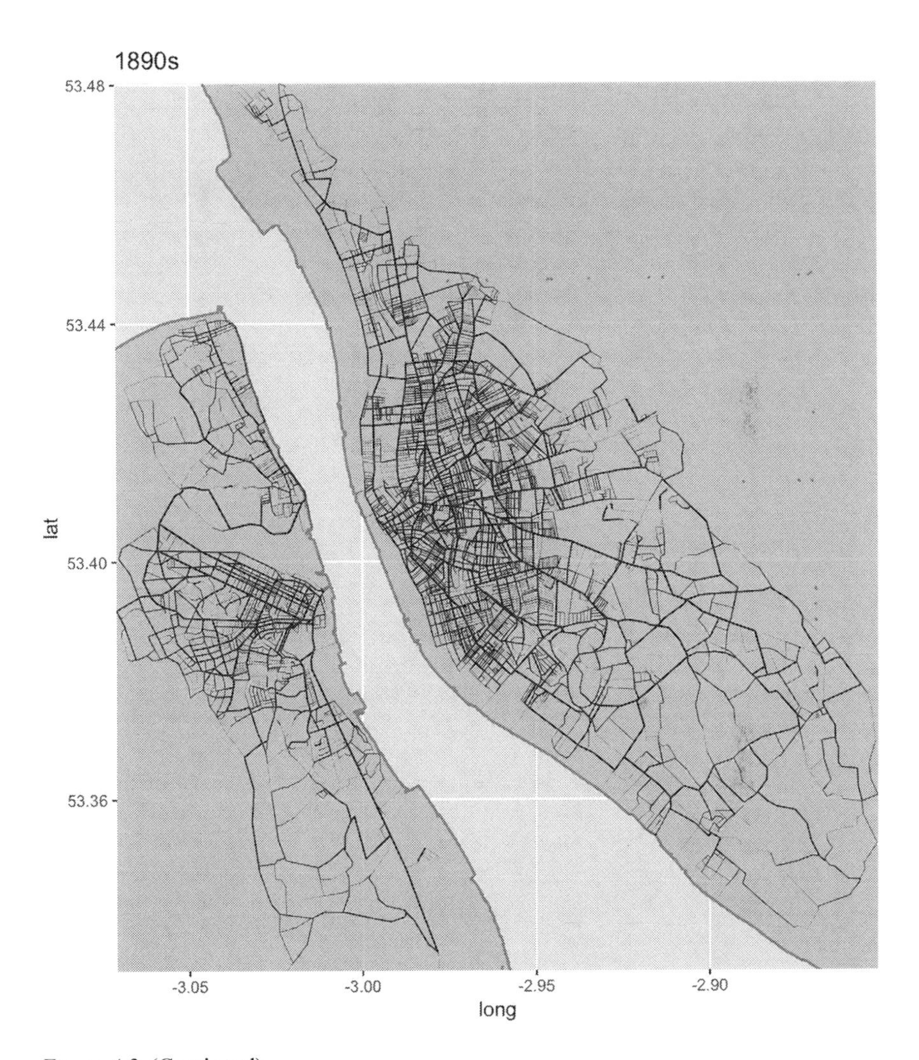

Figure 4.3 (Continued)

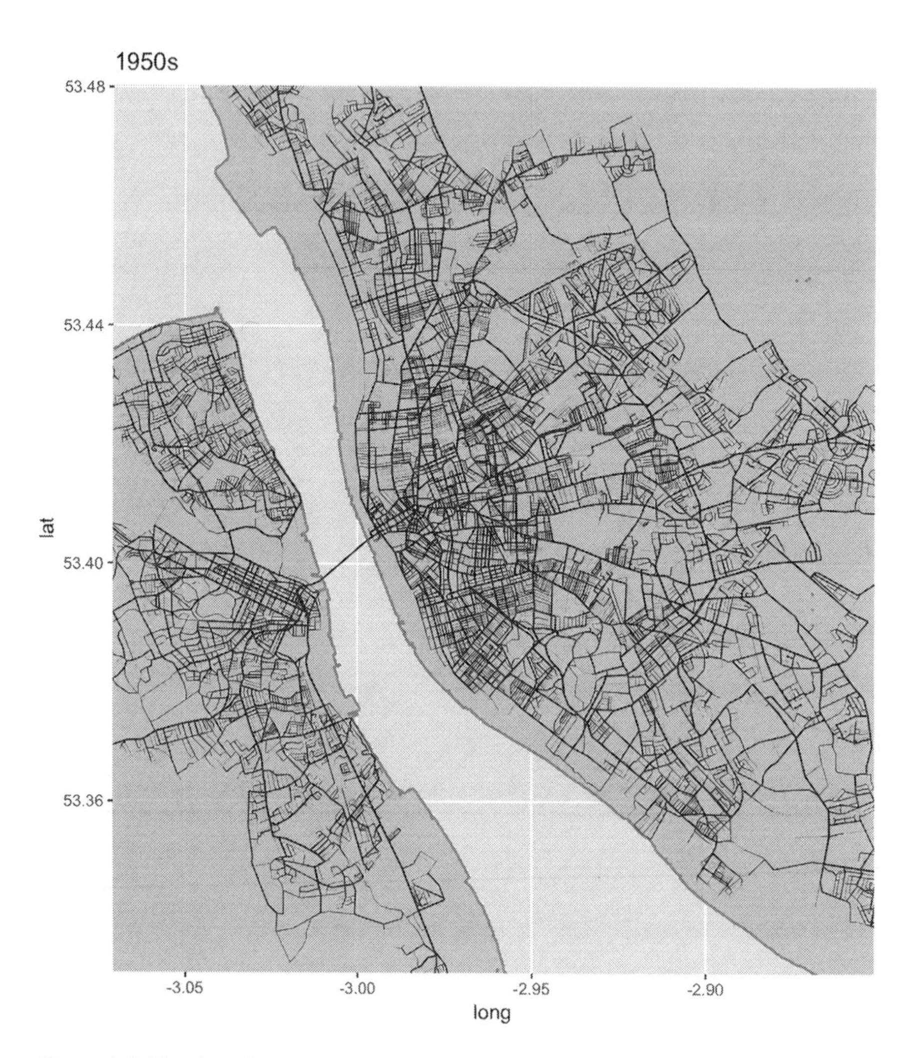

Figure 4.3 (Continued)

Figure 4.3 (Continued)

Princes Avenue, South Liverpool

Princes Avenue is a boulevard complex, built in the 1840s along an east-west axis over ancient parkland and gardens. The avenue connected the new metropolitan Princes Park to high-status residential areas. It was constructed as a dual carriageway that continues to play a major role in the urban formation of Liverpool's Toxteth area (Figure 4.4). By the 1850s, several local plots had been developed into streets with densified terraced housing, forming distinctive grid patterns. These sections presented overall low choice, compared with that of the burgeoning city centre, and were separated functionally from the wider network by the foreground structures of Upper Parliament Street (NAIN R5000).

The land surrounding Princes Avenue had been urbanized with high-density terraced-housing street sections (and given characteristic nicknames; see Figure 4.1, points 10–13). Movement potentials within the local street sections were based on generally low-movement structures of short, straight street segments (NACH 400, 2000). These were connected at the perpendicular to moderate choice-value segments, together bounded by street segments presenting high choice at wider scales (NACH R400, R2000 and R5000).

Separate to the high-density grids, the interstitial 'Granby Triangle' was the location of ostensibly higher-status housing, along with a cluster of landmark Protestant, Orthodox and Jewish places of worship. Princes Avenue afforded these householders improved accessibility to the city centre, to omnibus services, public parks, mercantile and cultural quarters (NAIN R800 and R2000). In contrast, road segments of the Welsh Streets and Holy Land seem to remain orientated to the riverside industries (NACH and NAIN, R400 and R2000).

By the 1950s Liverpool had undergone a second-wave expansion based around low-density suburban and peripheral developments, inter-connected by a radial complex that included the Queens Drive ring road (opened around 1920) and the Mersey 'Queensway' Tunnel (opened in 1934). Intensive wartime bombardment had led to several major disruptions to the urban fabric; nevertheless the Victorian urban network remained largely intact. The expansion of the network overall appears to have increased high choice values along Princes Avenue, as well as among its adjacent street-section areas (NACH R800, R2000, R5000). The array of rectilinear segments now inter-connected the street sub-networks (NACH R200), with movement flowing along Princes Avenue and into to the urban-scale centrality along Upper Parliament Street (NACH R5000). Together these provided consolidated accessibility from all street sub-networks to the city centre (NAIN R400, R800, R2000, R5000), although accessibility from Princes Avenue to the newly expanded network was apparently more limited (NAIN R5000, Rn).

The 1950s also mark a pivotal moment in Liverpool's economic and social decline. Granby's prosperity lasted from the city's economic zenith of the 1900s until the early 1960s. After this period, Granby experienced a rise in crime, which led to the widespread installation of bollard arrays to prevent kerb-crawling (SNAP, 1972),[1] which diminished choice and integration among the street network (NAIN and NACH R800, R2000).

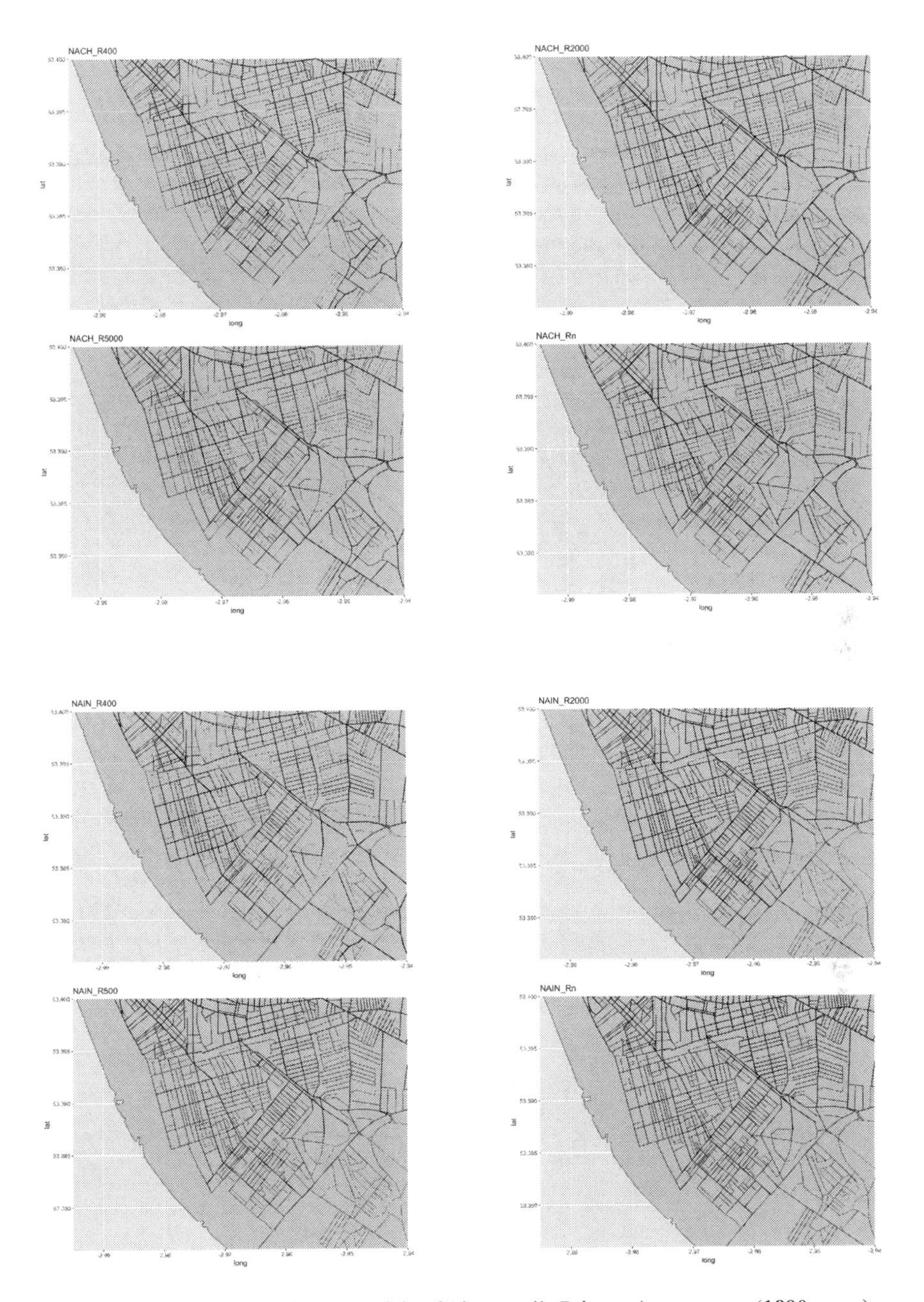

Figure 4.4 Arrays of Depthmap models of Liverpool's Princes Avenue area (1890s map).

This pattern of functional severance may have intensified a tendency towards spatial and demographic segregation, reflected in a concentration of non-white British populations within the Granby Triangle (HMSO, 1981). Relating to this, in 1981 a major riot broke out on Selbourne Street (an internally integrated segment within the Granby Triangle; NAIN R800), which led to the destruction or demolition of 70 buildings in the near vicinity. In the aftermath of these events, Liverpool City Council undertook the Urban Regeneration Strategy to develop additional low-density housing close to the city centre.

From contemporary observations, Princes Avenue presents high integration from Liverpool's southern periphery to the urban centre (NAIN Rn). Strikingly, the southern carriageway of Princes Avenue (conveying traffic into the city centre) now presents low choice (NACH R2000), compared to the generally high value of the 1950s. The avenue's segment lying perpendicular to the Welsh Streets appears to be a weak attractor for the local network (NAIN R400), which is perhaps a dynamic feedback from extremely high rates of vacancy along these streets. Also striking in the vicinity are the persistently high choice and integration values along the north-south axis of Lodge Lane, representing an emergent local- and city-scale centrality (NACH and NAIN, R400 and R5000).

Scotland Road, North Liverpool

Scotland Road is a major conduit situated on a natural elevation along a north-south axis, connecting Liverpool's urban centre to northern regional areas. It was developed as a highway in 1803, adjoined by an array of high-density terraced and court-yarded streets. Scotland Road (Figure 4.5) ran between distinctive areas of Vauxhall and Everton, which were associated with Protestant and Catholic migrants from within the British Isles.

Everton in the mid 19th century featured a localized 'ladder' pattern of short streets that afforded overall good accessibility to the wider urban network (NAIN R400, R800, R2000). These street sections were inter-connected via functional 'bridges' that traversed local- and city-scale urban sub-networks (NACH and NAIN, R400 and R2000), including economic centres of the mercantile quarter and riverside complexes.

The location of the North Hay Market, widely known as 'Paddy's Market', formed a major centrality for surrounding streets and the wider urban network (NAIN R400, R2000, R5000, Rn) (Figure 4.6). Highly significant in this dynamic was the perpendicular junction of Juvenal Street and St Anne Street, from where the market could be entered, forming a pivot of convergence for Liverpool's network globally (NAIN Rn and NACH Rn). The affordances for economic and social mobility brought about by these encounters are perhaps reflected in the innovative public housing established at close-by Summer Seat and Eldon Grove (1911–1912).

The contemporary map represents intensive redevelopment of the 1960s, including widespread demolition of the 'ladder' streets and construction of radial roads and the Mersey 'Kingsway' Tunnel complex (opened in 1971). This radial

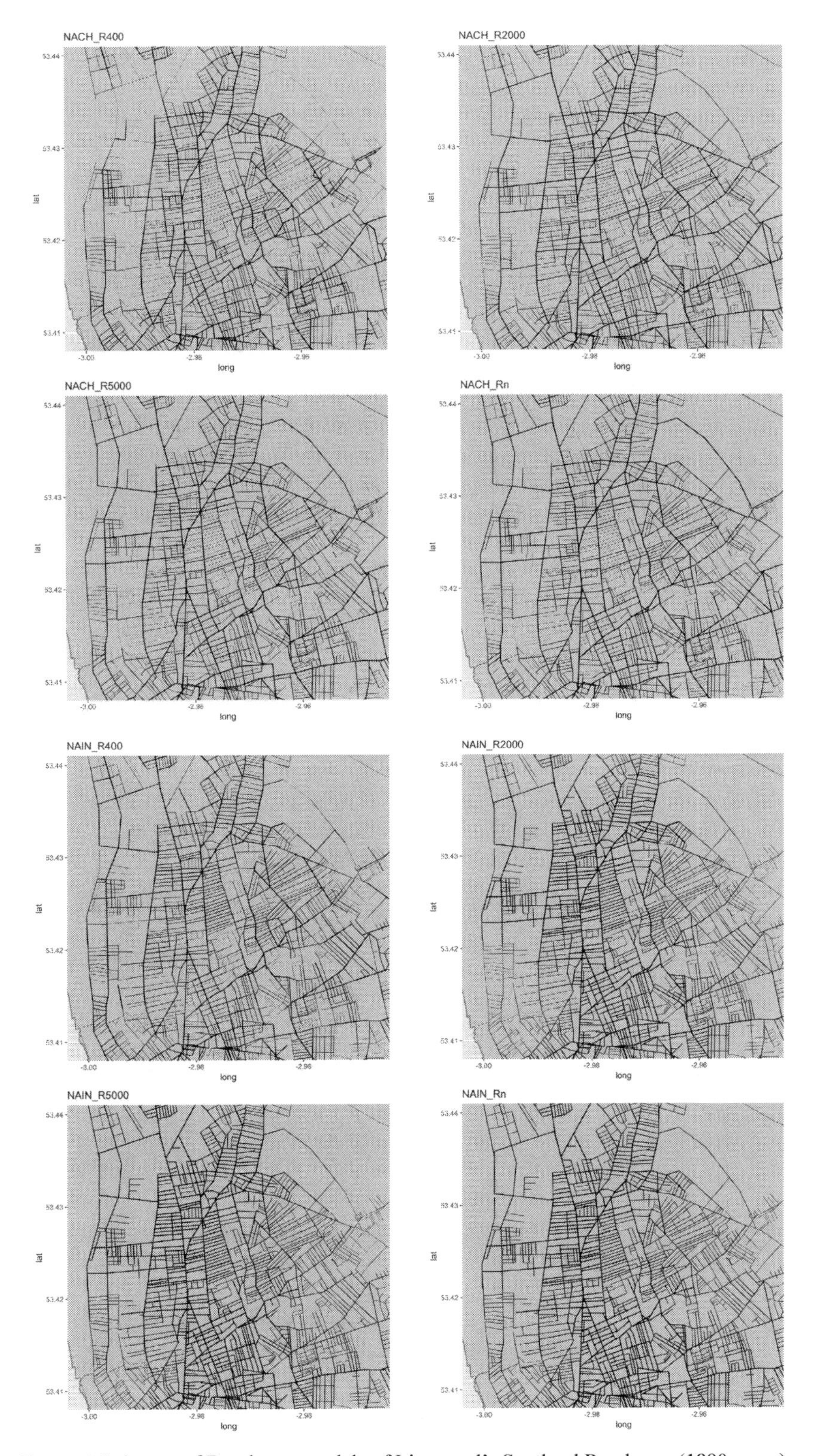

Figure 4.5 Arrays of Depthmap models of Liverpool's Scotland Road area (1890s map).

Figure 4.6 Depthmap array showing top quintile value ranges for NACH 400 and NAIN 5000, highlighting the ways in which local high choice around Paddy's Market may have coincided with a city-wide attractor.

expansion integrated the urban peripheries but formed prominent boundaries between the northern districts and wider network (NAIN R5000, Rn), and localized lacunae in the urban fabric. The location of Juvenal Street is now, sadly, a barren pedestrian bridge over the subterranean tunnel approach, bearing only moderate prominence as an attractor and conduit at all scales (NAIN AND NACH, R400, R2000, R5000 and Rn).

The sites of many of the 'ladder' streets have been redeveloped with low-density semi-enclosed complexes across north Liverpool. Together these form a loose cluster with overall low local integration (NAIN R400 and R2000), also demarcating a zone of among the highest multiple deprivations in the UK (Liverpool City Council, 2015). The low-density housing complexes have led to poor connectivity among the north-south axes of Scotland Road and Great Homer Street, with only St Anne Street affording direct access between the roads (NAIN R400 and R2000, NACH R2000). The overall poor connectivity between the north-south axes is perhaps suggestive of areas with weak local inter-connectivity and subject to movement from across the wider urban network.

Canning Place, Central Liverpool

Canning Place is the site of among Liverpool's earliest urban spatial structures and converges a historically persistent centrality at multiple network scales (Figure 4.7). From the 1850s, Canning Place formed a significant attractor within the city-wide network (NAIN R2000). At the local scale, Canning Place interconnected the mercantile quarter around Dale Street and the cultural quarter around St George's Hall (NAIN R400), and functionally integrated this sub-network with the Scotland Road axis via Whitechapel (NAIN R2000) and to regional routes (NAIN Rn).

The southern edge of Canning Place formed part of a boundary around the 'maritime quarter' sub-network, bisected diagonally by Paradise Street that traversed the 'mercantile' and 'cultural' quarters' sub-networks (NAIN R2000 and R5000). As such, the place's southern edge formed a powerful attractor and conduit for movement at all scales, converging along the city's foreground network (NAIN and NACH, R400, R2000 and R5000). Benefitting from this major, multi-dimensional centrality, Canning Place was the location for Liverpool's impressive Customs House.

By the 1950s, after intensive bombardment, the Customs House had been demolished along with most of the surrounding buildings. This led to a general decrease in movement potentials around the site (NACH and NAIN, R2000). However, the southern edge of Canning Place, which formed an axis with Hanover Street, remained functionally resilient to the damage in the surrounding network. It continued to serve as a conduit from Strand Street to the newly established radial routes into the city's expanded suburbs (NACH R2000, R5000, Rn).

Canning Place's 1970s redevelopment involved a complex of municipal buildings and Strand Street's redevelopment as a multiple-lane highway (Figure 4.8). Significant to the sub-network's connectivity with the wider network, the major

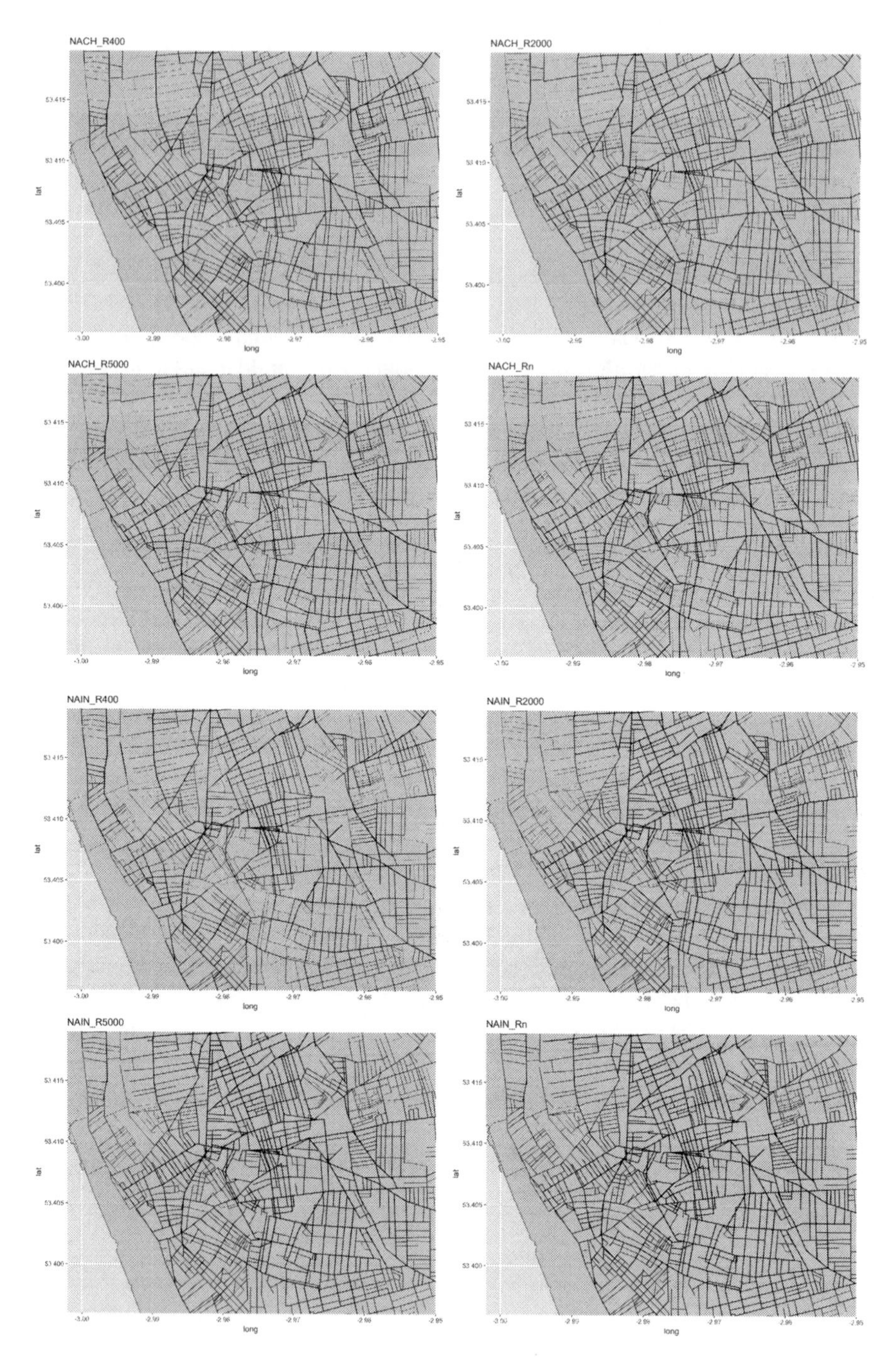

Figure 4.7 Arrays of Depthmap models of Liverpool's Canning Place area (1890s map).

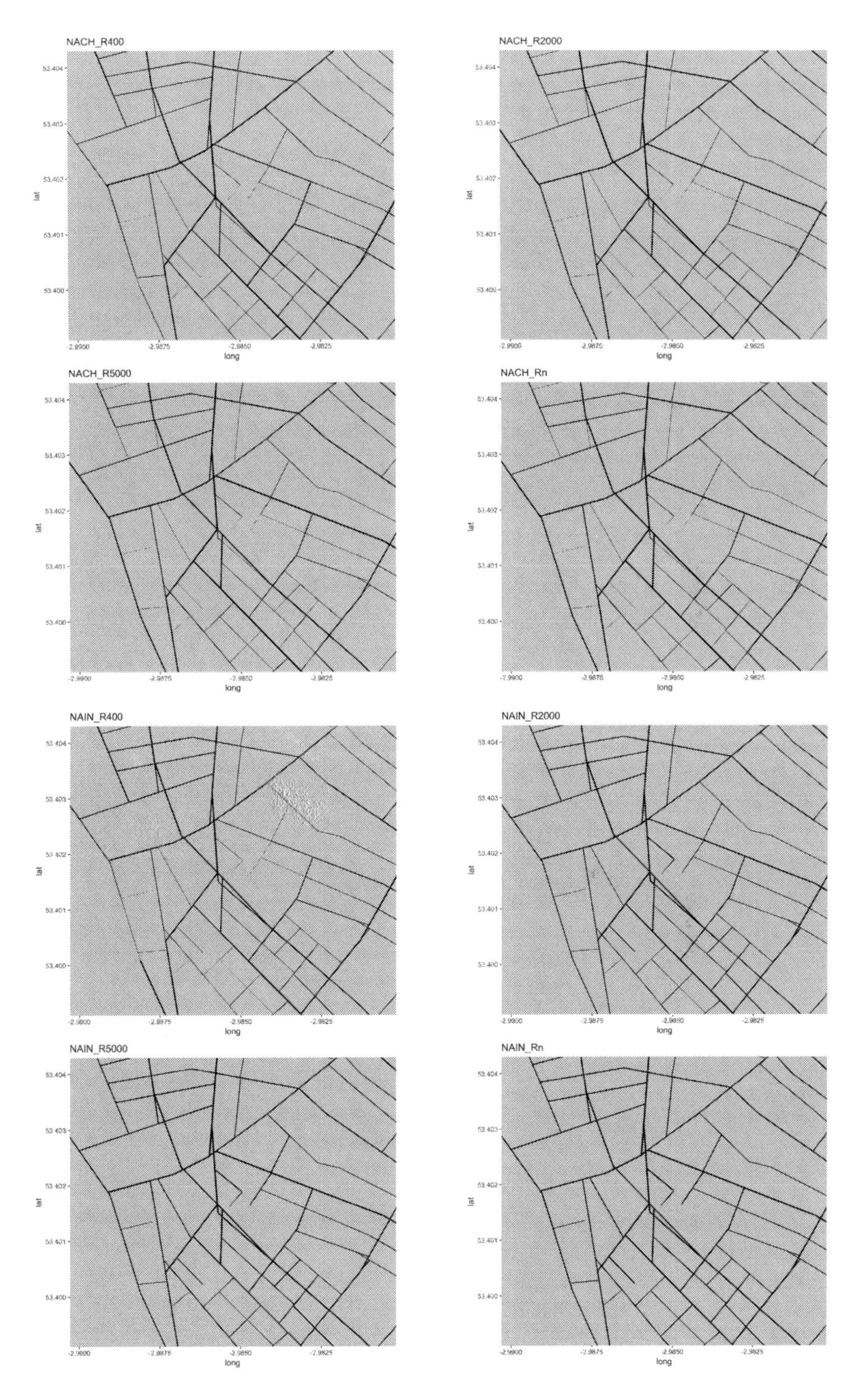

Figure 4.8 Arrays of Depthmap models of Liverpool's Canning Place area (1970s map; detailed).

axis of Upper Frederick Street that connected Canning Place with residential areas to the city's south (ultimately including the Princes Avenue area), had been devastated by wartime bombardment, yet was partially restored in the 1970s redevelopment. This served to increase choice values within the Canning Place network (NACH R400). However, a major housing development initiated by the City Council in 1984 severed the junction of Upper Frederick Street and Paradise Street.

Cable Street and Thomas Street had provided, until the wartime period, city-wide attractors and conduits that linked South John Street and Paradise Street (NAIN and NACH R2000). Notably, these structures formed part of a prewar sub-network (observable from the 1890s, NAIN R400). Their 1970s successor was a pedestrian pathway of sorts, passing through a car park and bus terminal. In spite of this architectural disruption, this pathway maintained its movement potential as a local attractor (NAIN R400). This functional persistence possibly speaks to a spatial logic for movement that has formed from the foreground structure of the Great George Street/Berry Street axis (forming the city centre's western edge), and its functional connectivity, via Seel Street/College Lane, with the natural flow of Paradise Street.

The area's redevelopment of 2004–2008 involved demolition of the 1970s architecture and pedestrianization of local thoroughfares. The development restored the earlier prominence within the network of Canning Place's northern edge, now called Thomas Steers Way (NACH and NAIN, Rn), to provide pedestrian access to a new retail area. What is today called Canning Place is formed from its erstwhile southern edge, comprising a walkway and roadway for the main bus terminal. The roadway has maintained its historical prominence at local and city-wide scales (NAIN R400 and R2000). However, its walkway that was once the location of the Customs House forecourt has become functionally disconnected from movements among Liverpool's foreground network and southern neighbourhoods (NAIN R400 and R2000).

Discussion

Our aim in using syntactic descriptions for making detailed diachronic observations of Liverpool's urban development has been to demonstrate how a 'normative' method for historical and spatial comparisons could reveal path-dependencies with the capability of underpinning urban images, including boundaries, thresholds, interfaces, gateways and bridges through which community identities are negotiated and controlled.

The Depthmap model arrays provided compelling observations of Princes Avenue's function as a boundary and threshold between the sub-networks to which it was inter-connected. On the basis of this research one can propose how populations occupying Granby, the Dickens Streets and Welsh Streets in the mid 19th century appeared to experience different orientations with respect to the wider network. Princes Avenue provided the major conduit for access to the metropolitan network, but with low inter-connectivity among the local networks. This suggests that Granby, the Dickens Streets and Welsh Streets were functionally separated and, we anticipate, socially segregated.

We discovered several examples of systemically confluent conduits and attractors that have, we argue, played various parts in community life at the different historical intervals we sampled. For example, Paddy's Market in the mid 19th century attracted localized movements, as we might expect from such a significant place in the urban landscape. The syntactical model also revealed how this local sub-network converged with through-movement structures at the wider city scale, possibly revealing its function as an interface between working-class communities to the north of the city and various populations from across the city. We also observed how the radial road developments of the late 20th century formed structural boundaries between these local sub-networks and the wider network, also reflected in their experiences of slum clearance, mass relocation and de-densification.

A historical threshold pattern is also detectable along Scotland Road that, in the mid to late 19th century and into the mid 20th century, provided a conduit into the city, but functioned as the dividing line between different local sub-networks. This major road provided accessibility to sub-networks proximal to riverside industries, meaning the various populations occupying distinctive (and segregated) areas encountered each other along its segments. We observed how functional 'bridges' around Scotland Road traversed residential and industrial areas. Their street corners and local pubs were likely to have been significant gateways for economic and social brokering, not least given the casual availability of work. These spatially 'trivial' yet socially important street segments warrant further analysis.

We also discovered the resilience in movement potential of segments relating to the Canning Place sub-network. The northern and southern edges of the place have persisted as a significant interfaces between various sub-networks including, in the late 19th century, those of the mercantile and cultural quarters and those of the socially mobile Princes Avenue quarters (though less so with regard to the lower working-class areas to the city's north). We included an overview of structural interventions of the 1970s, which weakened the functionality of Canning Place as a conduit and attractor within the local and wider networks; its northern edge in particular became demoted as a city-wide centrality. We also observed how redevelopment in the 2000s restored the northern edge's functionality within the local network, and promoted its southern edge within the wider network.

Conclusion

This chapter has offered a visualization method for making systematical observations of the historical growth of an urban street network that facilitates new forms of engagement with other sources of evidence and approaches to the study of urban history. This method has also produced evidence for the existence of normative 'images' of urban community formations in the historical street network.

Further work using, for example, contemporary demographic data, journalistic accounts and individual testimonies, as well as greater statistical validation, is required to establish the credibility of this hypothesis.

Note

1 See also https://asenseofplace.com/2015/04/25/the-bollards-of-liverpool-8-2/.

5 Community network toolkit I

Delineating and representing urban community formations poses a challenge of complexity to the design researcher. This complexity stems from the highly varied ways in which people make use of things in their local community spaces, which have evolved in various ways over time, as we saw in Chapter 4, to achieve their sense of place and location in relation to the wider urban landscape. For example, groups of people make use of local open spaces such as public parks, both to meet together and to determine boundaries between themselves and others who live close by. Similarly road networks provide a means of achieving connectivity between one place and another and also, from another point of view, the means of separating different 'neighbourhoods' in the wider area. For these reasons, a specific feature of a local urban area can be seen from different perspectives. A busy road junction can be both an intersection and a boundary. An open space can be both a place to socialize and a place for territoriality. The 'challenge' for the research described in this chapter was to find a way to identify and represent this relativistic complexity in a way that serves to enhance urban practices in planning and designing.

Network science provides a range of methods for representing relationships among individuals and groups based on membership profiles. Furthermore, statistical methods in this field allow these relationships to be measured in terms of associations, centralities, hierarchies and flows. These methods allow researchers to measure or infer aspects of social cohesion or segregation within the community under analysis. Network analysis also addresses a challenge in urban design research in that social network relationships can be explained in relation to spatial relationships among architectural or urban forms (cf. Hillier and Hanson, 1984; Grannis, 2009; Batty, 2013). In this chapter we take another distinctive perspective on the applications of network graph methods in urban domains, based on the theme of 'hierarchical design' (Batty, 2013; Alexander, 1971). This field of analysis supposes that the relational complexity of urban domains may be reflected in bipartite graphs that represent inter-connectivities among separate kinds of social or spatial data. These dual graph layers may be manipulated to generate a 'decision space' in which design options can be factorized, optimized and iterated.

The sections below outline some methods for preparing geo-spatial points data derived from urban community domains for hierarchical design applications. We are concerned in particular with digital modelling based on graph data structures, which provide a means for digital storage and retrieval. This chapter presents a range of models for the purposes of the current discussion, which are represented graphically. The programming code for these models has been written in R using the iGraph library, and is available via an open access repository[2].

Background

The urban community research activity, described in Chapter 3, yielded a rich set of volunteered data. This recorded the many ways in which community participants perceived their local community spaces, including structural and 'semantic' factors that affected their social interactions and sense of place. We outlined how the data were prepared using the open-source QGIS geographic information system. This GIS environment presented the data as points clusters, along with attribute tables that recorded the participants' profiles by age, gender and community group, the 'type' of community asset they selected, for example 'transport', 'open space', 'school' and so on, the specific name of the asset and any textual descriptions applied to that selection.

The next phase of the research involved aggregating the data to generate a community-level perspective. This level is an emergent property of the individual data sets, and we were able to depict this by aggregating and weighting the data clusters. Furthermore, we were able to observe the community-level data using different formats, including metric spatial and centrality graphs. Each of these afforded a different kind of observation, which together revealed otherwise hidden patterns in the community formation. The method for achieving this in an open-source programming environment posed some challenges, and this chapter opens with a practical guide to generating the graphs in different formats, before moving to a consideration of what they might represent about community formations.

Building a community network graph

Reading the following sections will equip readers with a general understanding of the methods we used to generate the network graphs. However, we also invite readers to follow the guidance steps and work through the practical examples. One caveat is that upgrades to the RStudio development environment and to the R libraries can result in compatibility issues when running the programmes. Readers who are entirely new to programming may benefit from working through some introductory tutorials in R and RStudio, which can be found online or in print form. If you are reading over the code and are familiar with languages other than R, note that R uses an assignment operator such as, for example:

```
DF1 <- data.frame('x'=B, 'y'=C)
```

This operator would send an instruction to create a new data frame object, called 'DF1', that binds vectors B and C into columns named 'x' and 'y', respectively. Another distinctive feature of R is the subset operator []. The position of the comma designates either columns or rows, respectively. For example: DF1[DF1$x == 1,] sends an instruction to subset data.frame 'DF1' by columns where values 'x' are equal to 1.

Setting up the programming environment

To set up the R programming environment, install an IDE (integrated development environment) such as the widely used RStudio on your computer and create a new R script. You can use your selected IDE to either input the R code line-by-line into the console's command line (designated here by the > prompt symbol, which is not to be included when you input the code), or else compile it in the script window and hit the Run command when you are ready to test your code. You will be working with spatial, network and digital image data, so you will need to install and attach the following libraries using this script:

```
> install.packages("sp", "rgeos", "maptools", "SDMTools",
"png", "iGraph", "dply", "qdapRegex")
```

Compile these in the console or script window:

```
> library(sp)
> library(regeos)
> library(dplyr)
> library(qdapRegex)
> library(maptools)
> library(igraph)
> library(png)
> library(SDMTools)
```

- sp provides classes and methods for manipulating spatial data.
- regeos provides an interface to a range of geometry manipulation tools.
- dplyr provides a comprehensive data manipulation grammar.
- qdapRegex provides a collection of regular expression tools for cleaning data.
- maptools allows you to upload GIS files, including the widely used .shp format, and apply coordinate references to the spatial data.
- iGraph is the key library for building, manipulating and visualizing network graphs. Other network graph R libraries are available (and, of course, available in other programming languages such as Python); readers are free to choose and assess the utilities of these libraries on their individual merits.
- png allows you to load digital image, or 'raster', data into the programming environment.
- SDMTools is useful in the present context for applying a scalebar to the plot.

Next, set the working directory to the computer folder in which you store your data:

```
> setwd("/my_data_path/my_data_folder")
```

Then create objects for the coordinate reference systems (CRS) for your spatial data, for example:

```
> ukgrid = "+init=epsg:27700"
> latlong = "+init=epsg:4326"
```

Now you can load your data. Once you have prepared your points data in a GIS environment, you should export this in .csv (comma-separated values) format. This .csv file will include the exact geo-coordinates for each of the points, and also allow you to clean your data based on the attributes tables. For example, you may wish to remove any rows with missing data, or to change string or numerical components of the unique IDs, which can make it easier to keep track of individual elements as you reformat your data.

```
> points_data <- read.csv("my_points_dataset.csv",
sep=",", header = TRUE, stringsAsFactors = FALSE)
```

Clean your data as necessary. For example, remove any missing data (that is, any data that has value NA):

```
> unique_ID <- unique_ID[-which(is.na(unique_ID))]
```

Or possibly reformat string or numeric components within the data:

```
> unique_ID <- rm_between(unique_ID, "-", "_",
extract=TRUE)
```

Finding asset groups

In this exercise, we are going to simplify the data by finding points clusters based on unique names (for example, all points with the name "Royal Park Street" or the code "RylPkSt"), and then find the central point within that cluster, which is called the cluster centroid. We achieve this by first creating a data structure that includes an equal number of columns and rows. In R this is a called a data frame, and is one of the key data structures in this programming environment (the other data structures include vector, list, array, and matrix). Data frames are particularly useful as they can contain any kind of data (including character strings, geo-coordinates or digital images), providing that the data vectors are of equal length.

We create the data frame (df) by extracting from the .csv file only those data that are needed for making the network graph. First we need the list of X and Y geo-coordinates (here they are longitude and latitude; but your coordinates may be in degrees), then the list of names, then the list of the types we have assigned to the selections.

```
> name <- as.character(points_data$specific_name)
> type <- as.character(points _data$urban_type)
```

```
> lon <- c(points_data$X)
> lat <- c(points _data$Y)
> df1 = data.frame(name, type, lon, lat)
```

We can look at the top lines of the data frame by running

```
> head(df1)
```

Name	*type*	*lon*	*lat*
0 RylPkSt	road	-2.955302	53.39030
1 RylPkRd	road	-2.958892	53.38669
2 RylPk	open space	-2.955185	53.38234
3 RylGds	open space	-2.957539	53.38093
4...			

Our aim is to plot the network based on geo-spatial and network graph formats. For the geo-spatial plot, you will need to extract the latitude and longitude coordinates vectors:

```
> lyt <- as.matrix(df1[,3:4])
```

Next we sort the data frame alphabetically by 'name', and then create a table called 'cnt' from 'df1', which applies a count of how many instances there are in which participants selected that particular urban asset:

```
> sort(df1$name)
> cnt <- plyr::ddply(df1,. (name), transform,
count = length(name))
```

We may want to simplify the data set by discounting assets that are deemed to be less significant to the community group. For example, among a group of 15 participants, Royal Park Street may have been selected by just two of the individuals, so we can say that this is less significant at the group level. We can create a subset of our data frame that includes only assets that were selected by the participants in greater than three instances:

```
> df2 <- subset(cnt, cnt$count > 3
```

We can now spatialize the subsetted data, stored in the df2 data frame. Firstly, create a new spatial points data frame (spdf1) based on df2:

```
> spdf1 <- data.frame(df2)
```

R structures the spatial data frame based on a set of headers and 'slots' (including 'data' and 'coords'). You can look inside the data slot by inputting > spdf1@data to the console's command line.

We can then attach the geo-coordinates, which are contained in the X and Y columns, and apply the coordinate reference system (CRS):

```
> coordinates(spdf1) <- ~X+Y
> proj4string(spdf1) <- CRS(latlong)
```

We need to find clusters of points within the spatial data. This is supported in R with the 'split' function that divides data into groups based on a specified identifier such as 'name', and can drop data that is not in a group:

```
> spdf2.gp <- split(spdf1, spdf1@data$name, drop=T)
```

Now the data has been split into groups, we need to store their general attributes in a new object 's', which includes only one instance of a point datum within each group. This is achieved with the logical construction data[!duplicated(data),]. In the R programming environment, an exclamation mark can be used to designate an obverse case (in other words, subset the data set where a vector is 'not duplicated'):

```
> s <- spdf1[!duplicated(spdf1@data$name),]
```

Returning to the grouped spatial data, our objective is to pinpoint the centroid of each points cluster as grouped under the 'name' attribute. We can achieve this by creating three new objects: 'x' that stores an integer value of the number of groups, 'h' that applies a spatial hull around each group and 'c' that applies a spatial centroid to each group (without need of an attribute identifier), and stores this as a spatial points data frame:

```
> x <- length(spdf2.gp)
> h <- sapply(spdf2.gp[1:x], gConvexHull, byid=FALSE,
id = NULL)
> c <- sapply(h, gCentroid, byid=FALSE, id = NULL)
```

Finding spatial centroids

Next we need to create a new data frame, 'cntrds.df' to contain the centroids vector. In this data frame, we set its length (number of rows) as being equal in number to that of the centroids data set 'c'. We can then use a 'for loop' to iterate over the data set and attach the centroids coordinates, with column names 'X' and 'Y':

```
> lc <- length(c)
> cntrds.df <- data.frame(matrix(ncol=2, nrow=lc))
> colnames(cntrds.df) <- c("X", "Y")
> for (i in 1:length(c)) {
cntrds.df[i,] <- c[[i]]@coords
}
```

Vectors that we stored in 's', including 'names', 'type' and 'count', can now be attached to the centroids data frame:

```
> cntrds.df$name <- s@data$name
> cntrds.df$type <- s@data$type
> cntrds.df$count <- s@data$count
```

Finally, when we come to plotting our centroids data, we may wish to adjust the size of each data point to reflect the total count of instances in which the asset it represents was selected. For example, if 'Royal Park' was selected by participants in 12 instances, then our point size could be 12 (or some other representative value). We might also wish to adjust the point size as a ratio of the total participants. For example, if there were 12 instances among a total of 400 data points (that is, 3% of the total), we could adjust the point size to 3 or 30.

Firstly, calculate the total number of data points with a unique identifier:

```
> sum.pt <- length(unique(unique_ID))
```

Next, add a new column 'perc' (percentage) to the centroids data frame to include a calculation of 'count' as a percentage of the total. Then round this percentage value to two decimal places:

```
for(i in length(cntrds.df)) {
cntrds.df$perc <- c(cntrds.df$count/100*sum.pt)
cntrds.df$perc <- round(cntrds.df$perc, digits = 2)
}
```

Finally, we spatialize the centroids data frame, following the same method outlined above:

```
> cntrds.spdf <- data.frame(cntrds.df)
> coordinates(cntrds.spdf) <- ~X+Y
> proj4string(cntrds.spdf) <- CRS(ukgrid)
```

Input > head(cntrds.spdf@data) to the console command line and the data frame, and the example data would look something like this:

Name	*type*	*lon*	*lat*	*perc*
0 J18	junction	-2.955302	53.39030	8
1 RylPk	open space	-2.958892	53.38669	15
2 ParkSt	road	-2.955185	53.38234	32
3 ParkAcd	school	-2.957539	53.38093	74
4 . . .				

Loading digital images (raster data)

The complexity of the urban community data means that the network graph we are aiming to visualize would benefit from schematic formatting. We can adjust vertex size and colour to reflect weights and categories within the data. Similarly, we can adjust line width and colour to represent degrees of connectivity among the vertices. We can also apply digital images to the vertices, which allows us to develop a bespoke symbology. This can be devised to reflect the specific characteristics of our community data set in profile. Working with R, we can attach rasters (digital images) to a data frame based conditionally on another value in a data frame column. These will provide us with two options for formatting the vertices: a percentage value and an iconographic raster.

Following our present example, we can attach images based on the 'type' column, so that a row storing element attributes, with a type category of 'open space', can be appended by an iconographic raster representing this kind of asset. This is presented illustratively below (in fact the data frame stores the raster as a numerical sequence, not as a viewable image):

	name	*type*	*lon*	*Lat*	*perc*	*icon*
1	RylPk	open space	-2.958892	53.38669	15	

We achieve this by creating a path to a repository of rasters that has been prepared to reflect the range of data by 'type' (for example, 'open space', 'junction', 'road', 'school' and so on):

```
> path2 <- "my_data/rasters_folder"
```

We then create a list of empty raster objects, equal in length to the total number of rasters in our folder. Keep adding imgType[n]=' ' objects until all the rasters you need have been included in the list:

```
rasters <- as.list(c(imgType1="' imgType2=",
imgType3=''))
```

We then add rasters to the list using the readPNG function (or else the readJPEG function, part of R's 'jpeg' library that you would need to install):

```
rasters$imgType1 <- readPNG(file.path(path2,"park.
png"))
rasters$imgType2 <- readPNG(file.path(path2,"road.
png"))
```

```
rasters$imgType3 <-
readPNG(file.path(path2,"junction.png"))
```

As above, keep adding imgType[n] based on the number of rasters you will need to use in your graph plot.

Next we need to create a look-up data frame, called 'lkp_df', that will support the conditional graph formatting and plotting. We are going to generate network data based on connections between points of origin and points of destination, designated as the 'from' and 'to' vectors, respectively. In this example, we are going to format the 'from' vectors only, so we create a data frame column (called 'from') based on the range of unique names in the 'df1' data frame:

```
> lkp_df <- data.frame("from" = unique(df1$name))
```

Then we create three new columns for the look-up data frame, including the 'type' of each name (for example, 'Royal Park' has type 'open space'). We achieve this by making use of R's match function to attach a 'type', and 'lat' and 'lon' coordinates, to each of the unique names, based on their match with names in 'df1':

```
> lkp_df$type <- df1$type[match(lkp_df$from,
df1$name)]
> lkp_df$X <- df1$lon[match(lkp_df$from, df1$name)]
> lkp_df$Y <- df1$lat[match(lkp_df$from, df1$name)]
```

Finally, we conditionally append the look-up data frame with an 'icon' column, which will contain the iconographic raster associated with the 'type' (for example, 'park.png' for 'open space'). This can be done using R's ifelse and grepl functions that apply the raster conditionally based on searches for matching character strings. Where no match is found, the raster field is designated as being empty, 'NA':

```
> lkp_df$image <- ifelse(grepl("park", lkp_df$type),
'imgType1',
ifelse(grepl("road", lkp_df$type), 'imgType2',
ifelse(grepl("junction", lkp_df$type), 'imgType3',
NA))))
```

Add each unique 'type' from your look-up data frame and each imgType[n] from your raster list. Add multiple brackets at the end of the function to close the each of the grepl opening brackets.

Adjusting vertex sizes

When plotting the vertices as iconographic rasters, we may also wish to adjust the vertex size to reflect a value range within the data, such as its percentage value; recall that this represents the ratio of instances in which an asset was selected to the total number of asset selections. We can append the look-up data frame with a 'size' vector. Firstly, create a temporary size column to contain the percentage values (based on string match as described above), and set this column as a concatenated vector:

```
> lkp_df$tmp.sz <- df1$perc[match(lkp_df$from,
df1$name)]
```

Use the ntile function to calculate the break-points for a quantile range of values (here we pinpoint five breaks, to create a quintile range):

```
> lkp_df$size -> ntile(lkp_df$tmp.sz, 5)
```

We can then use these quantile values to format vertex plotting. The sizes can, of course, be refined numerically depending on our formatting requirements.

Building a graph data frame

In the preceding sections we have gone through the steps needed to prepare spatial data for network graph formatting. In this section we outline a method for coding a network graph, and then using our spatial data to prepare this for plotting. R's iGraph library provides functionality for manually coding a network graph. This manual approach is useful as it allows us to make careful observations of the spatial points data, and to apply and test the data with different network configurations. Working with this library also allows us to plot the network data based on geo-coordinates, so that the vertices are plotted with respect to their metric distances, as they would appear on a cartographic map. The vertices can also be decoupled from their geo-coordinates, allowing their network relationships to be observed through different kinds of graph formats.

Firstly, create a new object 'gl' to store the results of a graph_from_literal function. For example, suppose we want to create a network graph involving several road intersections (marked as 'J' for junction plus a number), and to record whether the roads that connect them are one-way or have dual lanes. The graph_from_literal function has an intuitive way of recording these directional relationships using the symbols – (connection) and + (arrow) in combination. In the example below, the connection between J5 and J6 is one-directional, flowing from J5 to J6 but not the other way. The connection between J39 and J7 is two-directional, flowing from J39 to J7 and also from J7 to J39. Using this simple syntax, we can generate graphs at any level of complexity; here is a simple example:

```
> gl <- graph_from_literal(J5-+J6, J5+-J39, J39+-+J7,
J7-+J6, J7+-J8)
```

Inputting plot(gl) to the console command line will produce this simple network graph (Figure 5.1):

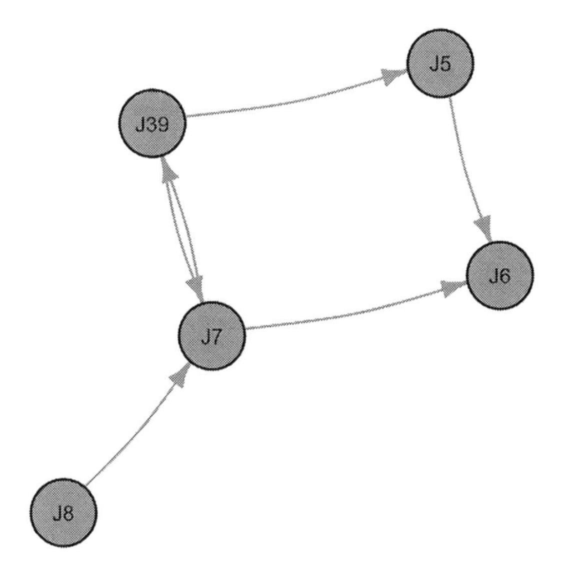

Figure 5.1 Simple network resulting from the graph code.

Once we have manually coded a network graph using this simple syntax, we then need to format the points of origin and destination as a graph data frame. We achieve this by extracting the edge lists – edges being the connective lines that join the vertices. These are organized in the iGraph environment as vector columns: the origins are in the first column, the destinations in the second column. We create two new vector objects, 'from' and 'to', assign to them the edge lists for origins and destinations respectively and bind them to a data frame, as below:

```
> from <- get.edgelist(gl)[,1]
> to <- get.edgelist(gl)[,2]
> df2 <- data.frame(from, to)
```

Now we can create the graph data frame, 'dg', which comprises the 'from'-'to' vectors of 'df2' and matches their vertices by name with those of the look-up data frame we created earlier:

```
> dg <- graph.data.frame(df2, vertices = lkp_df)
```

Note that the vertex names in your 'from'-'to' data frame must match exactly the vertex names in your look-up data frame. Any mismatched string will mean the graph data frame does not compile. The graph data frame stores vertex attributes within the V(graph_name)$[" "] structure, and edge data within E(graph_name)$[" "], as we see in the code snippet below.

The graph data frame does not yet contain the raster files from which the vertices will be plotted. We can bind these to the graph data frame by iterating over the data using the 'for loop' below. This loop function assigns the raster files by matching the vertex names in the graph and look-up data frames:

```
> for(i in V(dg)$name) {
imgtype <- lkp_df$image[lkp_df["from"]==i]
V(dg)[name==i]$raster <- rasters[imgtype]}
```

Formatting for metric observation

Once this code has been run, you can abbreviate the asset names to make them easier to read on the network plot:

```
> V(dg)$name <- abbreviate(V(dg)$name, 4,
strict = TRUE)
```

The graph can now be formatted for plotting based on geo-spatial coordinates, as in this example:

```
> par(mfrow=c(1,1))
> par(mar=c(1,1,1,1))
> par(bg="white")
> plot.igraph(dg, layout=lyt, vertex.shape="raster",
margin=.1,
vertex.size1=V(dg)$size,vertex.size2=V(dg)$size,
vertex.label = V(dg)$name, vertex.label.color="gray35",
vertex.label.dist=-1, vertex.label.degree=-pi/2,
vertex.label.family="Arial",vertex.label.cex=0.6,
edge.width=0.4,edge.arrow.size=0.05, edge.
color="darkorange", rescale=T)
```

The rescale function fits the data to your window by normalizing the geo-coordinates as metric distances. You can add a title and control its position in the plot window by adjusting its 'line' value:

```
> title("Metric distance", cex.main=0.8, col.main =
"grey10", line = -7)
```

As this graph represents metric distances among the vertices, as they would appear on a cartographic map, working with RStudio you can also include a scalebar to represent distances in metres or kilometres. To position this, you can input locator() to the command line and use the cross-target cursor to pinpoint the best location for your spacebar on the plot window (hit the escape key to reveal your selected coordinates). You can then insert the xy coordinates manually and input the code. Here is an example (you will need to insert the xy coordinates from your own graph plot):

```
> Scalebar(x=-0.7758591, y=-0.8733493, distance=1)
```

Finally, we can add a legend to show the range of vertex sizes included in the graph, based on an approximation of the exact sizes and their actual value ranges. Create a list of the unique vertex sizes included in the graph (in this case, there are three unique values):

```
> u <- sort(unique(lkp_df$size))
```

Then create an empty data frame, and use a 'for loop' to bind the value ranges to this:

```
> rng <- data.frame()
> for(i in 1:length(u)) {
s <- subset(lkp_df, lkp_df$size == u[i])
rng <- rbind(rng, range(s$tmp.sz))}
> colnames(rng) <- c("min", "max")
```

Next, create a list to contain the value ranges; here the list items are named 'kl' 'kj', and so on, and the value ranges vertex sizes (which are approximates of the vertices plotted to the graph):

```
kl <- list()
kj <- list()
for(i in 1:length(u)) {
kl[[i]] <- paste(", rng[i,][1], '-', rng[i,][2], '%',
sep='')
kj[[i]] <- i+0.5*0.9
}
```

Now run this code to add a legend to the graph plot, which includes example xy coordinates:

```
> legend(x=-1.090902, y=-0.7330582, ncol = 3,
adj=c(1.2,2.5),
c(unlist(kl)), pch=22, pt.cex=c(1.0, 2.5, 3.8),
bty = "n", title="Vertex size: % range",
text.font=3, text.width=0.08, cex=0.9,
box.lty=2, box.lwd=2)
```

Running the plot.igraph, spacebar and legend functions will produce a metric distance network graph, with vertices formatted as iconographic rasters (Figure 5.2).

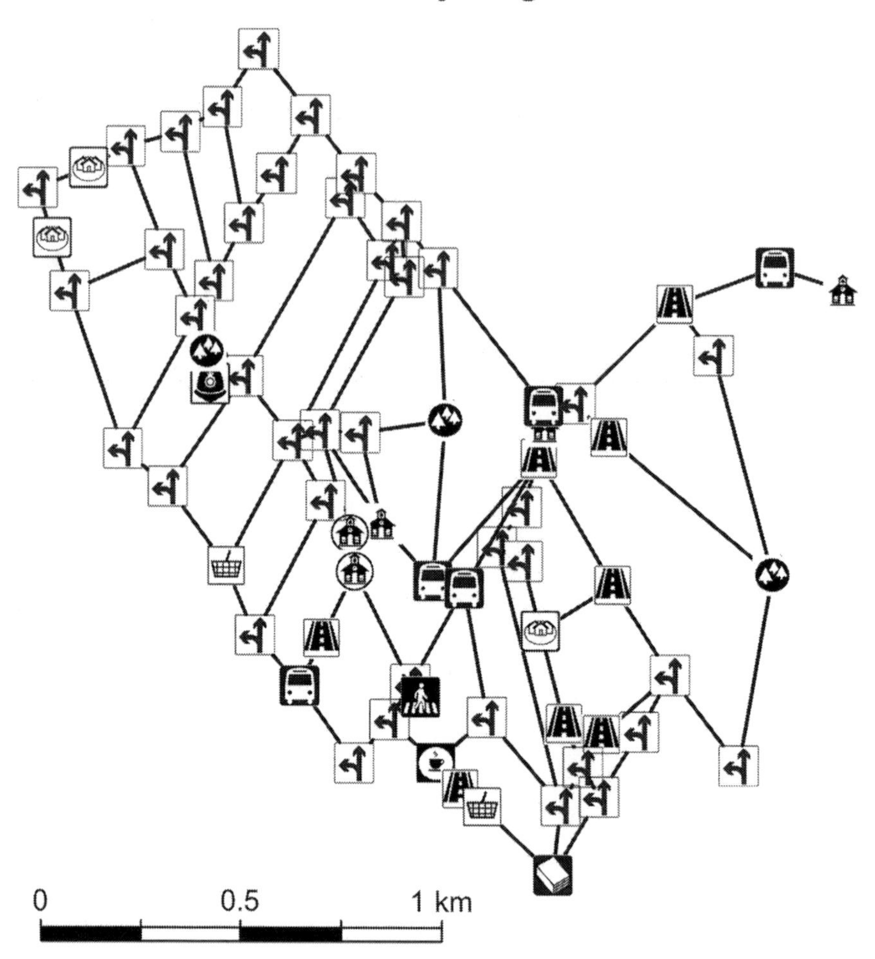

Figure 5.2 Geo-located network graph (metric scale shown) with iconographic vertices.

In the next section we outline some methods by which vertex sizes may be adjusted in relation to their 'weighting' within the community data set, as in Figure 5.3.

This graph is based on anonymized data from the study outlined in Chapter 3. The iconographic vertex rasters represent the range of assets selected by the community participants. The sum of instances in which the assets were selected is represented by the vertex sizes, and we can observe how certain roads, parks, schools, a café, a library and a police station were deemed to be major assets by the community participants. Most of the junctions recorded here were deemed to be less significant.

This community weighted graph represents the sets of community assets as they would appear on a cartographic map. We are also interested in understanding the patterns of connectivity among community assets as a functional network (decoupled from geographic space), and we explore this theme in the next section.

Formatting for network analysis

The community weighted graph outlined in the sections above showed us one kind of connectivity among the vertices. This was based on distance relationships among the assets selected by the community participants. This is very useful where we wish to understand how the community relates to the assets in terms of their 'real' proximity. Networks function in various ways, and we can format the network of assets based on a range formats, to reveal different kinds of relationships among them. Centrality is a measure of the relative importance of vertices or edges within the network as a whole. Vertices or edges with the highest centrality values are the most important in terms of getting from one part of the network to another, or else connecting one part of the network to other parts. In this section we will go through the steps to generate a graph that highlights patterns of centrality in the network data.

Betweenness centrality

In this section we will focus on the centrality value of betweenness, which is based on the total count of shortest paths that pass through a vertex. Betweenness quantifies the number of instances in which a vertex or edge serves as a bridge among the network's shortest paths. An edge with a high betweenness value will usually bridge more than two edges within the network. The greater the network complexity, the higher the number of paths the vertex will bridge. Betweenness is a very useful quantifier for our present example, as it helps to reveal relationships within the network of community assets that are not easy to see at the geographic level, as we outline below.

To observe the properties of betweenness, we create a new 'analytical' graph object based on the graph data frame we created in the section above:

```
> ang <- as.undirected(dg, mode = "collapse")
```

We then calculate the betweenness values of 'ang' by applying a function from the iGraph library. The resulting values represent the sum of shortest paths that are inter-connected via each vertex:

```
> btwns <- igraph::betweenness(ang, directed = FALSE)
```

We then standardize the number of pairs of vertices being joined by the 'bridging' vertex (the vertex count does not include the bridging vertex):[1]

```
> btwns_std <- btwns/((vcount(ang) - 1) * (vcount(ang)
- 2) / 2)
```

You can view these standardized betweenness values in your console by binding them to a data frame:

```
node_btwns <- data.frame("btwns" = btwns, "btwns_
std" = btwns_std)
```

We can also apply to the graph a measure of edge betweenness:

```
> edge_btwns <- igraph::edge_betweenness(ang,
directed = FALSE)
```

The graph can now be formatted with a well-known network layout (which we outline in the following section), for plotting based on betweenness centrality, as in this example:

```
> plot.igraph(ang, layout = layout.kamada.kawai,
vertex.label = V(ang)$name, vertex.shape = "raster",
margin=.2, vertex.size=btwns * 0.03,vertex.size2=btwns
* 0.03,
vertex.color = V(ang)$color, vertex.label.
family="Arial",
vertex.frame.color = "grey30",
vertex.label.dist=-1, vertex.label.degree=-pi/2,
vertex.label.color = "grey35", vertex.label.cex = 0.6,
edge_btwns * 0.01, edge.arrow.size = 0.5,
edge.color = "darkorange", edge.lty = E(ang)$lty)
> title("Betweenness", cex.main=0.8, col.
main="grey10", line = -3)
```

Running this plot.igraph function will produce a network graph, with vertices formatted as iconographic rasters and their sizes adjusted in relation to their betweenness values. Edge widths are also adjusted to reflect their betweenness values, and we can see in Figure 5.3 the vertices and their connecting edges with highest betweenness.

Comparing the community weighted graph with the betweenness graph, we can observe some striking differences. Significantly, the betweenness graph reveals how the community assets selected by participants in most instances, such as the public parks and schools, bear very low betweenness values. This means that the assets that are most significant to the community as a whole are not evenly inter-accessible across the network. The vertices that support greatest inter-accessibility among the edges are two junctions, 'J29' and 'J1', which were both selected by just one community participant.

The mean of the edge betweenness range ($n = 70$) is 94.9 ± 65.3. A table plot of the top ten highest-value edges, in descending order (Table 5.1), reveals how over 418 shortest paths are linked via the J1-J29 configuration. This is a far higher betweenness value than all other edges.

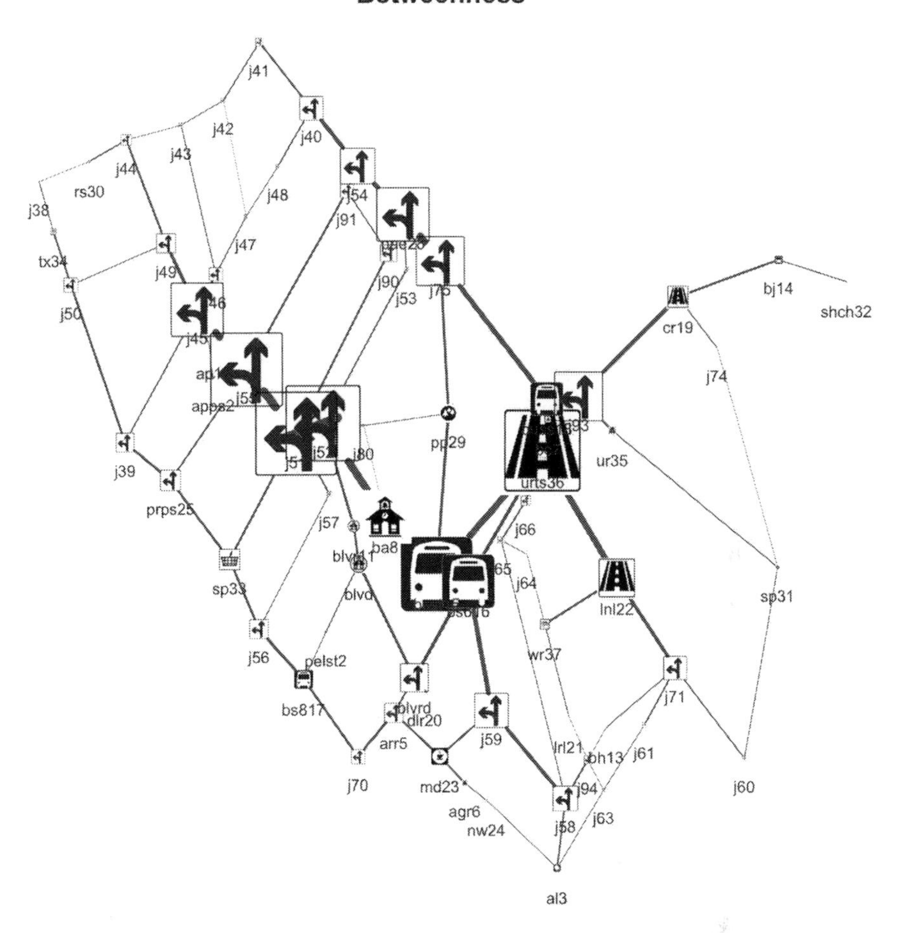

Figure 5.3 Network graph with edges weighted by betweenness and vertex size adjusted for community weighting.

Table 5.1 Betweenness values for the sample's network edges

edge	*betweenness*
J1|J29	418.4286
J7|J8	235.0000
J28|J29	217.1714
J1|J9	214.8833
J27|J29	210.4381
J3|Pr_P	195.2619
J8|J9	171.8333
J3|B_R_	166.3952
J1|Pr_P	164.4119
J23|B_R_	154.5619

Closeness centrality

In this section, we outline applications of closeness centrality, which is based on the total lengths of the shortest paths between one vertex and all others in the network. As we are dealing with a metric spatial layout (that is, based on distances between vertices as they would appear on a cartographic map), closeness provides a useful quantification of how close one vertex is to all others in the network. We can generate a closeness graph, for example:

```
> clsns <- igraph::closeness(dg, mode = "total")
> clsns_std <- clsns/(vcount(dg) - 1)
> node_clsns <- data.frame(vertex= V(ang)$name,
closeness = clsns,
clsns_std = clsns_std) %>%
tibble::rownames_to_column()
> dc <- node_clsns %>%
arrange(-closeness) %>%
.[1:10,]
```

The table of closeness values reveals that the range varies only minimally. For this reason, adjusting the vertex size would reveal few details about their rank within the graph, as shown in Table 5.2.

Instead, we will look at the rank order by decoupling the vertices from their (normalized) geo-coordinates, and showing them arranged in a centre-to-periphery format. We can achieve this by applying a force-directed graph format (here, we use the well-known Kamada Kawai layout), in which centrality is modelled as a form of 'energy' as it exerts influence across the network. The vertices with highest centrality are shown at the centre, and those with the least at the edges, as shown in Figure 5.4.

In this example, we can see how the 'public park' vertices named 'Pr_P' and 'Sf_P' form sub-group clusters within the network. This in an interesting observation, as these features were not ranked highly in the betweenness centrality graph, outlined in the section above. However, they were ranked very highly by

Table 5.2 Closeness values for the sample's network vertices

vertex	*closeness*	*clsns_std*
J1	0.001848429	3.242858e-05
J29	0.001845018	3.236874e-05
J27	0.001824818	3.201434e-05
Pr_P	0.001821494	3.195603e-05
J28	0.001779359	3.121683e-05
J3	0.001773050	3.110613e-05
P_A_	0.001773050	3.110613e-05
J15	0.001769912	3.105108e-05
J9	0.001766784	3.099622e-05
Sf_P	0.001754386	3.077870e-05

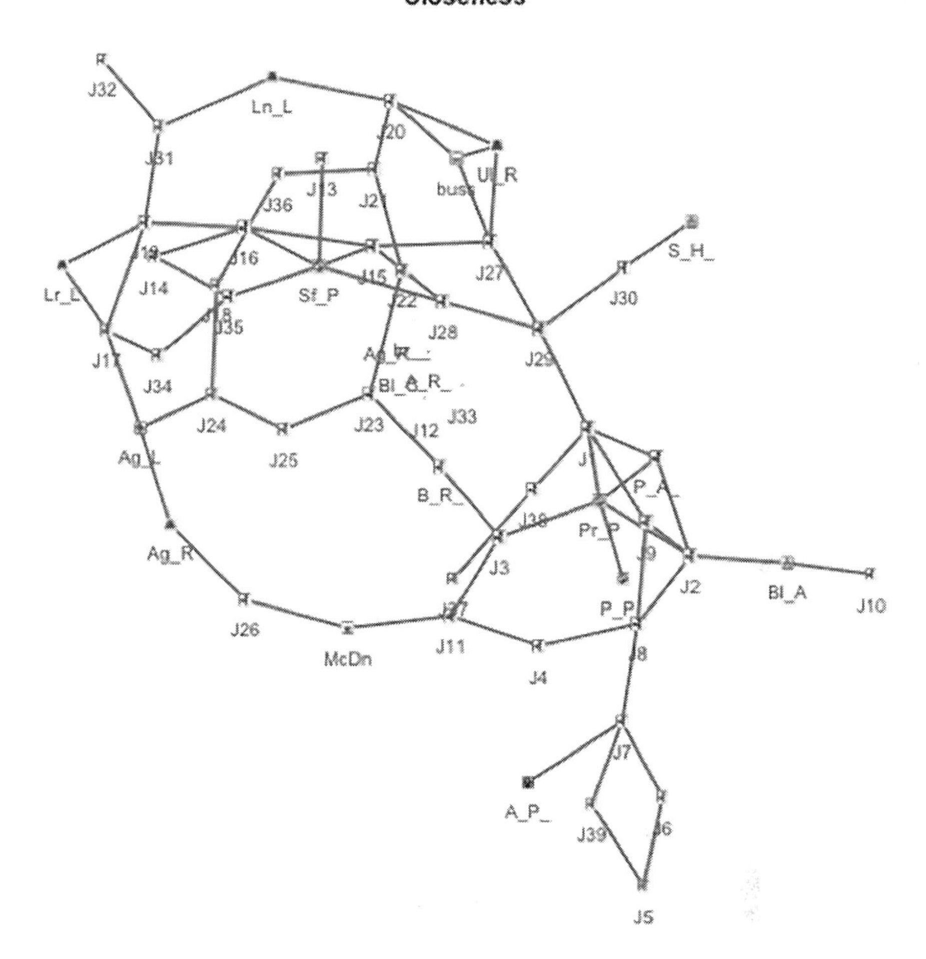

Figure 5.4 Network graph with vertex sizes adjusted for closeness.

the community participants, as shown in the community weighted graph, which we also outlined above. It appears in this example that the closeness graph, which quantifies total path lengths between the vertices, is closer to the participants' own sense of importance of relationships among their community assets.

Observing these kinds of outliers and differences can inspire questions for design research in urban domains, which could be framed around a set of questions, for example:

- Is the importance of the J29-J1 configuration somehow subsumed into the everyday life of the community? If so, how can this be highlighted in a community decision-making process?

- Could it be that the community assets that are accessible via J29 and J1 are important, but that these vertex 'bridges' are not meaningful as such to community life?
- Does the J1-J29 configuration coincide with a real urban structure? For example, a specific road? If so, can we reveal this road segment's urban-network value using space syntax?
- Can the 'real' importance of this configuration be improved for the benefit of the community as a whole? For example, can the network be reconfigured to improve inter-accessibility with the J29-J1 configuration?

Conclusion

In this chapter we introduced some methods for generating network graphs that can be formatted with geo-spatial and graph layouts. In other words, the geo-located vertices can be decoupled from their coordinates in order to observe their network centralities based exclusively on network structure. We also introduced a method for plotting vertices as rasters, which allows us to encode network graphics with iconographic schemata. In Chapter 6 we develop this methodological theme for applying weights to vertices and generate directed graphs of attractors for community formations. In Chapter 7 we outline more advanced ways in which these spatial relationships might be represented as iconographic schemata that serve to describe *cognitive and historical knowledge* (MacEachren, 2004, pp. 185–190), in which community formations are embedded.

Notes

1 Further details about this and other standard graph processes mentioned in this chapter are available in general textbooks on network theory, including van Steenen (2010).
2 Available to download at www.routledge.com/9781138652057

6 Community network toolkit II

In the previous chapter, we outlined a method for representing complex community relations with local spatial assets based on force-directed graph formats. This provided one useful way of showing how a community makes use of a range of assets in its local area. This also provided a 'weighting' to the representations of relationships based on the number of instances in which the community members had selected the asset within the participatory research exercise. In this chapter we develop this graph-based approach to community spatial visualization by making use of the community members' descriptions of the assets they selected. The participatory research exercise from which the sample data was derived invited community participants to select any spatial structures within their local areas, which often included assets such as roads, junctions, parks and schools, based on how these helped or hindered local people in their everyday community interactions. For example, a road might help people in travelling to meet together, but might also provide a boundary from crossing from one urban area to another.

Given the manifold complexity of community spatial assets, the researchers also invited the participants to provide a brief text description of assets they deemed to be particularly significant. The resulting body of text descriptions was highly mixed in quality and quantity, and impelled the researchers to apply a text-analysis method to draw out themes from across the data. We applied a 'keywords-in-context' approach, from which we could then configure graph-based visualizations of the community spatial assets. The advantage of this approach is that the graph can be configured to reflect the relational complexity of community spaces from the perspectives of the community members. Moreover, common perspectives merge around community sub-groups, which are themselves latent properties of the community inter-relationships. For example, a sub-group of community members emerges among those who share a set of experiences relating to crossing a busy road. The individual members in this example might present differing profiles, including age-group, gender or demographic status. Yet their descriptions of their experiences in the urban landscape can, under analysis, yield commonalities in experiential presentations that are not apparent in their categorical presentations. In other words, working with text descriptions can tell us a great deal about how a community comes together based on their shared experiences, rather than on the 'accidents' of their situations in life.

Setting up the programming environment

In this section, we repeat some of the steps for setting up your programming environment presented in Chapter 5, to allow readers to treat these materials as stand-alone guidance. Readers may also choose to skip the repeated steps without missing the main objectives. Once you have set up the RStudio environment, as introduced in Chapter 5, you will be working with spatial, text-based and image data, so you will need to install and attach the following libraries using this script:

```
> install.packagesl("tm", "qdapRegex","wordcloud",
"SnowballC, "sp", "stringr", "sampSurf")
```

Compile these in the console or script window:

```
> library("tm")
> library("qdapRegex")
> library("wordcloud")
> library("SnowballC)
> library("sp")
> library("stringr")
> library("sampSurf")
```

- tm is a text mining library for processing, analyzing and finding patterns in statements or bodies of text.
- qdapRegex provides a collection of regular expression tools that can be used for cleaning data.
- Wordcloud provides tools for formatting and representing patterns in text.
- SnowballC provides tools for finding connective patterns in text, based on 'stemming' from one keyword to other related keywords.
- Stringr supports manipulation of text, including finding and matching.
- sp provides classes and methods for manipulating spatial data.
- sampSurf includes a function for constructing bounding boxes and polygons, which is useful for clipping spatial data.

Next, set the working directory to the computer folder in which you store your data:

```
> setwd("/my_data_path/my_data_folder")
```

Then create objects for the coordinate reference systems (CRS) for your spatial data, for example:

```
> ukgrid = "+init=epsg:27700"
> latlong = "+init=epsg:4326"
```

Now you can load your data. Once you have prepared your points data in a GIS environment, you should export this in .csv (comma-separated values) format.

This .csv file will include the exact geo-coordinates for each of the points, and also allow you to clean your data based on the attributes tables. For example, you may wish to remove any rows with missing data, or to change string or numerical components of the unique IDs, which can make it easier to keep track of individual elements as you reformat your data.

```
> points_data <- read.csv("my_points_dataset.csv",
sep=",", header = TRUE, stringsAsFactors = FALSE)
```

Clean your data as necessary. For example, remove any missing data (that is, any data that has value NA):

```
> unique_ID <- unique_ID[-which(is.na(unique_ID))]
```

Or, possibly reformat string or numeric components within the data:

```
> unique_ID <- rm_between(unique_ID, "-", "_",
extract=TRUE)
```

Finding keyword groups

The participants of a recent research project into urban community formations were invited to write down their thoughts and feelings about spatial assets in their local areas, including roads, buildings and open spaces, that they deemed to be significant in either helping or hindering people in coming together. The participants' text descriptions varied very widely in quality, from brief phrases to fully articulated sentences. For this reason, the researcher decided to extract lists of keywords from the text descriptions that related to the major assets selected by the participants. The keywords were then counted and concatenated as a list, and this list was joined to the respective asset spatial points objects.

To achieve this, we firstly load the .csv file in which our participatory data has been collated:

```
> data <- read.csv("my_data.csv", sep=",",
header = TRUE, stringsAsFactors = FALSE)
```

We extract a manageable subset based on names of the participant groups:

```
ds <- d[grepl("(name_group1)|(name_group2)",
d$element_ID),]
```

We then create a data frame to contain the points data for each of the spatial assets included in the data subset:

```
X <- c(ds$X)
Y <- c(ds$Y)
```

```
name <- as.character(ds$strct_spec)
type <- as.character(ds$strct_type)
df1 = data.frame(X,Y,name,type)
sort(df1$name)
```

Count the number of instances in which each asset was selected by participants. Perhaps it is helpful to think of this as the 'votes' for each asset, which participants cast based on either positive or negative connotations:

```
cnt <- plyr::ddply(df1,. (name), transform,
count = length(name))
```

Create a new data frame to include only those assets that were selected in three or more instances:

```
df2 <- subset(cnt, cnt$count > = 3)
```

Then spatialize the data frame holding the points data for plotting. Note that you also need to set the Coordinate Reference System (CRS) to your local system. In this example, we use the reference for British National Grid:

```
> crs = "+init=epsg:27700"
> spdf1 <- data.frame(df2)
> coordinates(spdf1) <- ~X+Y
> proj4string(spdf1) <- CRS(crs)
```

We can now split the spatial points data frame into separate groups, based on the names of the assets selected by the participants. We also create a new spatial object to include only the top row of each group; this will contain all the data we need for this task:

```
> spdf2.gp <- split(spdf1, spdf1@data$name, drop=T)
> s <- spdf1[!duplicated(spdf1@data$name),]
```

Next we generate each group's convex hulls and calculate their centroids:

```
> x <- length(spdf2.gp)
> h <- sapply(spdf2.gp[1:x], gConvexHull, byid=FALSE,
id = NULL)
> c <- sapply(h, gCentroid, byid=FALSE, id = NULL)
```

We then create a spatial object to include all data relevant to the assets. We start this by binding the geo-coordinates of the points data centroids to a data frame, using a 'for loop' to iterate over the list of convex hulls. After this is complete, we join the columns of vectors that we need:

```
> lc <- length(c)
> data_cntrds.df <- data.frame(matrix(ncol = 2,
nrow=lc))
> colnames(data_cntrds.df) <- c("X", "Y")
> for (i in 1:length(c)) {
data_cntrds.df[i,] <- c[[i]]@coords
}
> data_cntrds.df$name <- as.character(s@data$name)
> data_cntrds.df$count <- s@data$count
> data_cntrds.df$type <- s@data$type
> data_cntrds.df$cols <- s@data$cols
```

When plotting the spatial data as a network, we may wish to format the graph based on the number of instances in which an asset was selected ('voted for'). For example, we can adjust the sizes of vertices based on the number of votes as a percentage of all the votes cast:

```
> sum.pt <- length(unique(partpt_ID))
> for(i in length(data_cntrds.df)) {
data_cntrds.df$perc <- c(data_cntrds.df$count/100*sum.pt)
data_cntrds.df$perc <- round(data_cntrds.df$perc,
digits = 2)
}
```

We are also going to manually write a network graph, so it will also be useful to apply abbreviated code names to vertices, which help to avoid errors and reduce the effort involved in writing (explained in full below). First, create a new object to contain our data:

```
n <- data_cntrds.spdf@data$name
```

Then use a string-splitting function to extract the first letter of each word in the asset names:

```
> nx <- lapply(n, function(x) {
x1 <- str_split(str_trim(x), " ")
paste0(unlist(lapply(x1, str_sub,1,1)), collapse="")
})
```

Then format and clean these abbreviations, leaving us with a vector of brief, fairly uniform code-words for the assets, which we bind to our data frame:

```
> nx <- tolower(nx)
> nx <- sub("[^A-Za-z]+", "", nx)
> data_cntrds.spdf@data$code_id <- paste0(nx, >
rownames(data_cntrds.spdf@data))
```

Processing community data

Now that we have created our data frame to contain data about the selected spatial assets, we return to our collated data to extract and prepare the text descriptions provided by the participants.

At this stage we can remove any rows on the data frame that do no contain text description (using the !is.na, meaning 'not nothing', function) and ensure the column names are well defined, as we need these to select and match data (your column names may differ from the example provided here):

```
> ds_text <- subset(ds, !is.na(ds$description))
> colnames(ds_text) <- c("X", "Y", "id", "age",
"gender", "icon", "type", "name", "text")
```

We now need to extract keywords from the text descriptions, and there are some simple ways of achieving this. Firstly, just remove any words with three or fewer letters:

```
> ds_text$text <- rm_nchar_words(ds_text$text, "0,3")
```

Next, extract the keywords of the specific spatial assets, firstly by splitting the ds_text data set into a list itemized by the unique asset names:

```
> nm_ls <- split(ds_text, ds_text$name, drop=T)
```

A 'for loop' can then iterate the nm_ls list to remove common words that are not usually keywords, which cleans the text generally, and then to 'tokenize' the strings of words so as to isolate the keywords. Keywords are then counted, arranged in descending order and added to a list, ds_text_ls:

```
> ds_text_ls <- list()
> for(i in 1:length(nm_ls)) {
df3 <- nm_ls[[i]]
df3$text <- tolower(df3$text)
df3$text <- tm::removeNumbers(df3$text)
df3$text <- str_replace_all(df3$text, "  ", "")
df3$text <- str_replace_all(df3$text, pattern =
"[[:punct:]]", " ")
df3$text <- tm::removeWords(x = df3$text,
stopwords(kind =
"SMART"))
df4 <- subset(df3, df3$text != "")
df5 <- df4 %>% unnest_tokens(word, text)
word_freq <-
```

```
ddply(df5,.(word),transform,count=length(word))
td <- data.frame(word_freq)
td <- td[order(-td$count),]
td <- td[!duplicated(td$word),]
type <- td$Category
cat <- unique(td$Category)
td <- na.omit(td)
ds_text_ls[[i]] <- td
}
```

We can remove any of the ds_text_ls items that have not received keywords:

```
> ds_text_ls <- ds_text_ls[sapply(ds_text_ls, nrow) > 0]
```

Finally, for this part of the process, we create a new data frame to include the asset name and a vector of its keywords and their counts:

```
> dj1 <- data.frame()
> for(i in 1:length(ds_text_ls)) {
n <- as.character(ds_text_ls[[i]]$name[1])
f <- ds_text_ls[[i]]$count
w <- ds_text_ls[[i]]$word
tst <- data.frame(n,w,f)
tst$m <- paste(tst$w, ': ', tst$f, sep='')
j <- list(tst$m)
dj <- data.frame(paste(j))
dj$name <- n
dj <- dj[c(2,1)]
colnames(dj) <- c("name", "kywds")
dj1 <- rbind(dj1, dj)
}
```

We then bind the keywords and counts to a spatial points data frame, which we will use to process and plot the spatial data. We change the column name from 'name' to 'node', which will make it easier to bind data in the next stage:

```
> nodes <- data_cntrds.spdf
> nodes@data <- left_join(nodes@data, dj1, by="name")
> colnames(nodes@data)[7] <- "node"
```

We will return to working with keywords in the section, "Ranking the vertices", below, where we will use a library of words arranged into categories of 'positive' and 'negative' connotations. This will allow us to apply a score to the community keywords.

Finally, in this stage we can create a bounding box as a regular polygon around the spatial limits of our set of points objects ('nodes'), which will be useful for clipping geo-spatial data, especially road-network vector layers:

```
> bx <- bbox(nodes)
> rownames(bx) <- c("x", "y")
> sp.bx <- bboxToPoly(bx)
> proj4string(sp.bx) <- CRS(latlong)
```

Constructing the undirected graph

We are now ready to construct an undirected graph of our community network. Plotting this graph with geo-spatial coordinates requires us to construct a 'look up' data frame called 'meta' that will hold the geo-coordinates and any attribute data we wish to include in the graph plot:

```
> meta <-
data.frame("node"=c(as.character(nodes@data$node)),
"lon"=c(nodes@coords[,1]),
"lat"=c(nodes@coords[,2]))
> meta$node <- as.character(meta$node)
```

We can manually construct a geo-located network graph based on any spatial points data set. Figure 6.1 shows an example of a sample of spatial points plotted over a road network.

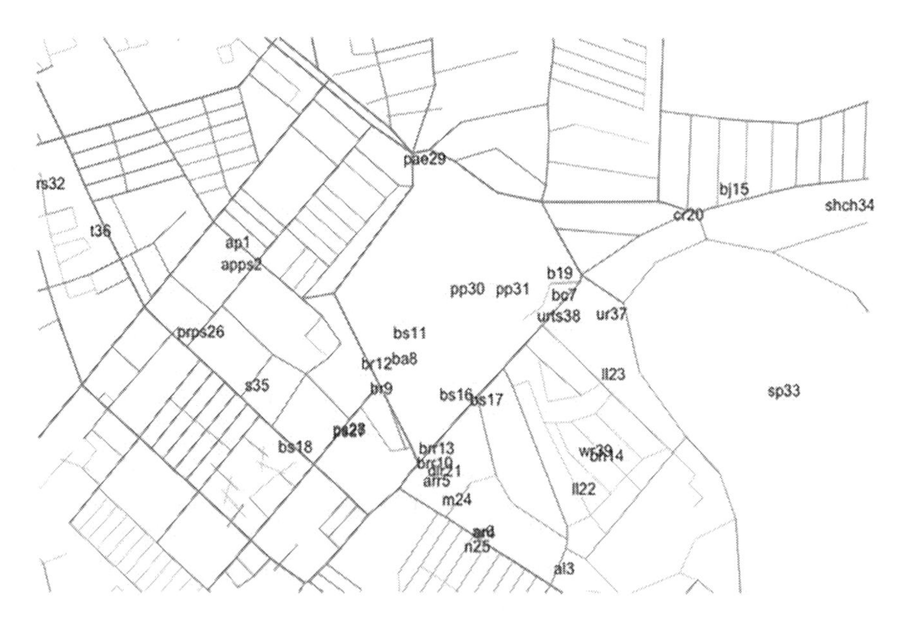

Figure 6.1 Plot of selected community assets over sample road network.

Table 6.1 Manual record of connections between assets (vertex list)

node1	*node2*	*node3*	*node4*	*node5*
pp25	pp26	j62	ba7	j38
pp26	b15	bc6		
bc6	urts33	j45		
b15	j45			
j45	j44	ur32		
j62	pae24	j61		
j61	na69			
na69	j60			
j60	j38	j47		
j38	br9			
br9	br8	ba7		
br8	brr10			
brr10	dlr17	j40		
dlr17	arr4			
arr4	j40	m20		
m20	ar5			
ar5	n21			
n21	j41			
j41	al2			
al2	ll18			

Following this example, construct the graph in a comma-separated format using any spreadsheet application. Start with any spatial point and observe all the points that it should connect to in the graph. This often involves a point at a road intersection being connected to point at the next intersection along the road. Decide for yourself how your graph should be constructed. Table 6.1 is an example of the graph construction showing the first 20 of the spatial points.[1]

The task now is to transpose this table into long-form format. This allows us to generate a data frame to include vertex pairs that are connected, albeit without direction. We will describe a way to create a directed graph in the section, 'Constructing the directed graph', below.

Now we are going to use our data frame of node airs to construct a geo-located undirected graph. Firstly, import the .csv file in which you wrote your graph. Here I use the .csv file name 'spatial_graph.csv', but your file name may be different to this:

```
> spatial_graph <-
read.csv("/my_dir/my_folder/spatial_graph.csv")
```

Then create an empty data frame to contain the nodes:

```
> node_pairs <- data.frame()
```

Now write a 'for loop' to iterate over the rows in the spatial graph object. This will take the first point name in each row, which will be the 'from' vertex, and then the rest of the point names in that row, which will be the 'to' vertices, and clean the data. The loop will then transpose the rows into columns, by repeating the 'from' vertex to the same length as the 'to' vertices. Taking the first row of our sample .csv above, this would result in a long-form construction such as:

```
pp25 pp26
pp25 j62
pp25 ba7
pp25 j38
```

The long-form 'from' and 'to' vertices are then bound to a data frame and appended to the node_pairs data frame:

```
> for (i in 1:nrow(spatial_graph)) {
nd2 <-
data.frame(spatial_graph[i,][2:length(spatial_
graph[i,])])
nd2 <- apply(nd2, 2, function(x) gsub("^$|^ $", NA, x))
nd2 <- na.omit(nd2)
f <- as.character(spatial_graph$node1[i]) #'from' vector
length of 'to' vector
nd1 <- rep(f, length(nd2))
df1 <- data.frame(nd1, nd2)
colnames(df1) <- c("node1", "node2")
row.names(df1) <- NULL
node_pairs <- rbind(node_pairs, df1)
}
```

We can now construct our 'look-up' data frame by extracting all of the unique vertex names from the node_pairs data frame, and matching the data contained in the meta data frame:

```
> l1 <- data.frame("node"=unique(node_pairs$node1))
> l2 <- data.frame("node"=unique(node_pairs$node2))
> lkp_df <- rbind(l1, l2)
> lkp_df <- data.frame("node"=lkp_df[!duplicated(lkp_
df),])
> lkp_df$node <- as.character(lkp_df$node)
> lkp_df$X <- meta$lon[match(lkp_df$node, meta$node)]
> lkp_df$Y <- meta$lat[match(lkp_df$node, meta$node)]
```

The layout coordinates for plotting the graph can be contained in a separate matrix, called 'lyt':

```
> lyt <- as.matrix(lkp_df[,2:3])
```

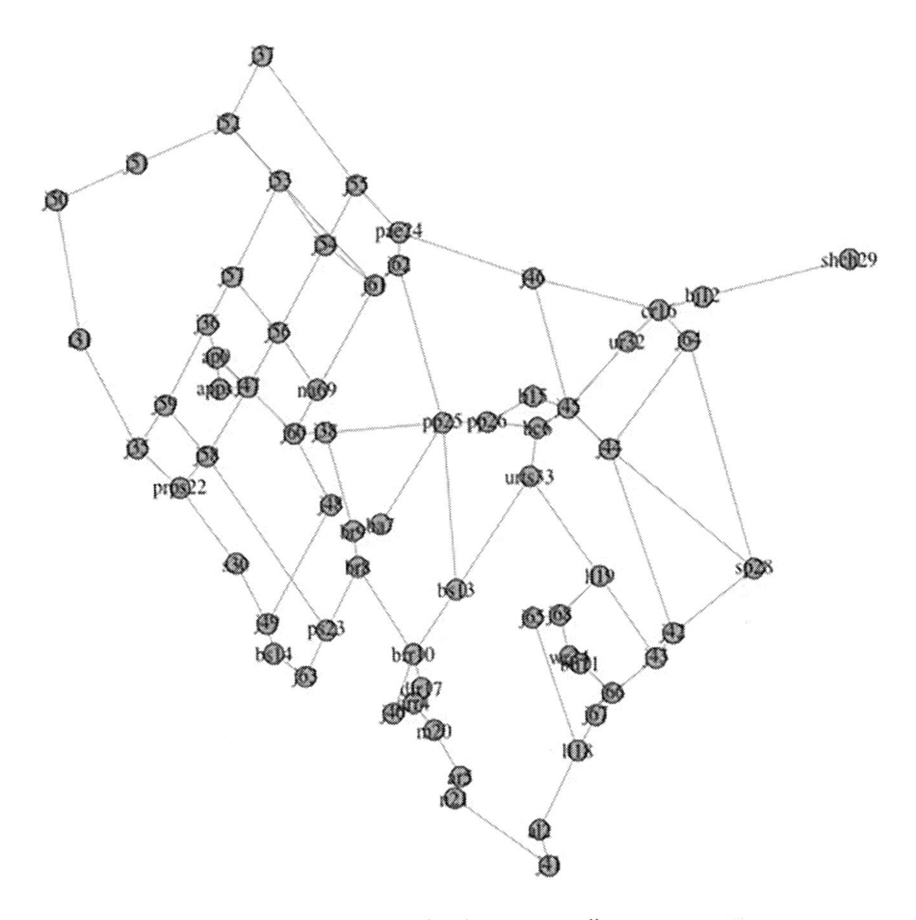

Figure 6.2 Connections among vertices fixed to geo-coordinates.

We can now generate our graph, using a graph data frame structure from the iGraph library, and generate a simple test plot (Figure 6.2):

```
> dfg <- data.frame(node_pairs$node1, node_
pairs$node2)
> g <- graph.data.frame(dfg, vertices = lkp_df,
directed = F)
> plot(g, layout=lyt, vertex.size=5)
```

This graph shows a set of network relationships among the spatial points included in our data set. The benefit of this is that we can now analyze these relationships based on different kinds of distance, as well as on geographic proximity. These other kinds of distance might relate to routes through the network that involve all or some of the vertices, or on the inter-accessibility among points, showing the 'cost' involved in getting from one vertex to another. Network relationships can be calculated and explored using well-known measurements such as betweenness and closeness, and the centrality of vertices (measuring their connectivity) can be calculated using degree and eigenvector values.

As our focus is on geo-located networks, we need to consider how network centralities are also affected by geographic parameters. We can achieve this by adding 'weight' to vertices based on the availability of geographic information, such as proximity or community significance. This significance might be positive for community life, as in the example of street corners along dense road networks that afford social exchanges, or they may be negative, as in the locations of environmental hazards, or experiences of crime and social exclusion. The vertices may be also associated with vector attributes relating to, for example, topography, demographics or to aspects of community life.

To look at the local significance of vertices, as represented by their attributes, we can apply a community-detection algorithm to the network, and then apply local weights across the modular communities ('clusters'). There are several widely used community-detection algorithms, including fast greedy, leading eigenvalue and Louvain, and spinglass is most appropriate to our 'from-to' graph data structure as it finds those pairs of vertices that are naturally bound together and builds up these basic relationships towards a set of modular communities. We create a new data object to contain the modular communities generated from the spinglass algorithm:

```
> clst <- cluster_spinglass(g)
> plot(clst, g, vertex.size=5, layout=lyt)
```

This results in the network community-detection graph shown in Figure 6.3.

Ranking the vertices

In this section we describe a method to score the vertices based on their keyword attributes, which we extracted and appended to the community data set. We can achieve this by deciding whether a keyword has a 'positive' or a 'negative' connotation from the point of view of community welfare. Examples of keywords with positive connotations might include *friends*, *chatting* and *enjoyable*, while those with negative connotations might include *gangs*, *strangers* and *dangerous*. It is also possible that keywords have ambiguous connotations, such as the word *people*, or are bivariate, in the sense that they have both positive and negative connotations. In these instances, their connotation depends on the context in which the keyword appears. We will deal with this issue below.

Create lists of keywords with positive and negative connotations, such as those included in the following example:

```
> pos_words <- c("friends", "mates", "cafes",
"festivals", "love", "house",
```

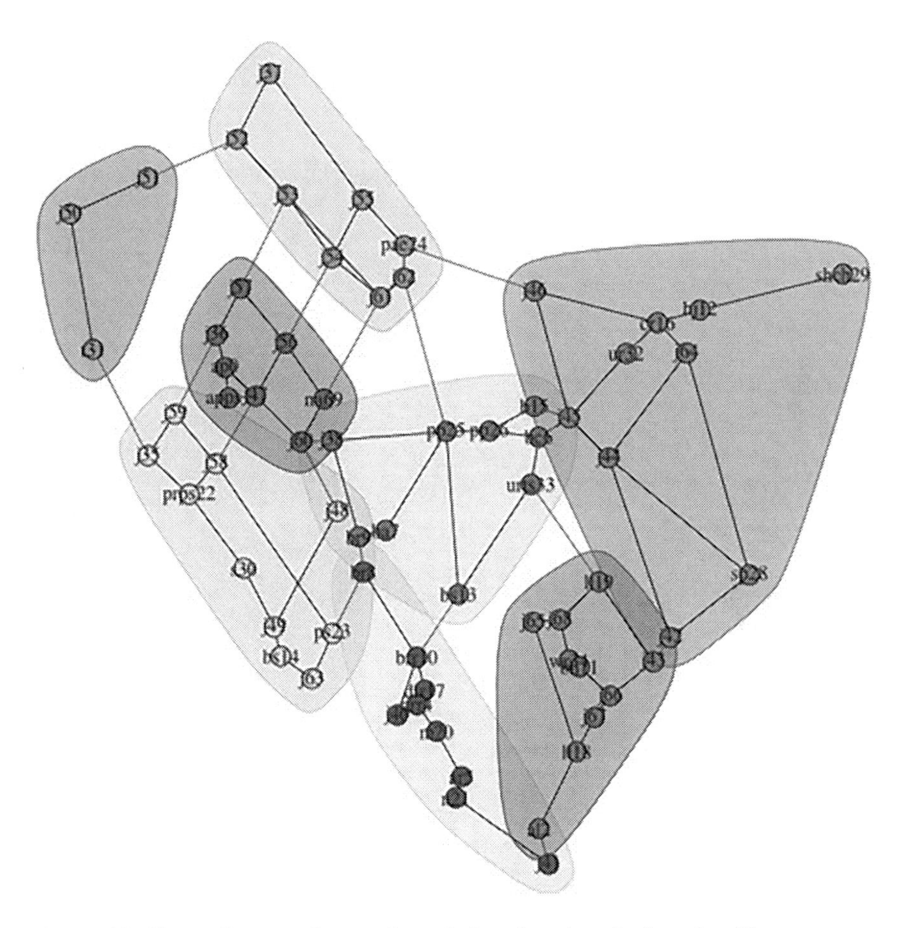

Figure 6.3 Cluster diagram of network graph, based on the spinglass algorithm.

```
"home", "people", "visit", "friendly", "family",
"safe", "learn", "happy", "socialise", "families",
"visit", "nice", "teachers", "social", "talk")
> neg_words <- c("avoid", "scary", "unsafe", "gangs",
"shocking", "people", "hate", "scared", "danger",
"dangerous", "night")
```

You can include any keywords in either category that seem appropriate to your own community data set. We can now use these connotative categories to apply a score to each of the keywords in our community data set. We may wish to highlight community spatial assets that are particularly positive for community welfare, so we apply a high score of 1.0 to positive keywords, a low score of 0.1 to negative keywords and a medium score of 0.5 to ambiguous or bivariate keywords. We will also increment the score based on the number of vertices in each cluster, as a proportion (%) of all the vertices in the data set. This means that vertices benefit from being in larger clusters, where their 'influence' is more easily shared. Once the incremented scores have been applied, these can be ranked. In this example we rank by quintile (five breaks).

Firstly, create a object to contain the sum of all counted vertices:

```
> sk <-
sum(spdf_text@data$count[!is.na(spdf_text@
data$count)])
```

Next, a 'for loop' extracts each of the keywords and the count of their instances, determines whether they can be matched to a connotative category, applies a score and binds these to a new data frame called node_score. Then the loop calculates the sums of the scores for each of the ranks. This will allow us, in the following loop function, to adjust the highest scores based on the co-occurrence of low and medium scores. In other words, we will 'temper' the high scores where there are lower scores also associated with that vertex. To help in following this function some notes have been added, annotated with the # prefix:

```
> node_score <- data.frame()
> node_list <- list()
#iterate over each of the nodes in the data set
> for(i in 1:length(spdf_text)) {
nd <- spdf_text[i,]
#extract keywords and counts; bind to df
n <- nd$node
k <- nd@data$kywds
k <- c(rm_between(k, '"', '"', extract=TRUE))
k <- as.factor(k)
k1 <- sub(':.*', ", k)
k2 <- as.numeric(sub('.*:', ", k))
kdf <- data.frame("node"=n, "word"=k1,"count"=k2)
#kdf$qunt_cnt <- ntile(kdf$count, 5) #quintile rank
#node rank scores
r <- data.frame()
#sub-loop to apply scores
for(i in 1:nrow(kdf)) {
x <- if(kdf$word[i] %in% pos_words) {1.0}
else if(kdf$word[i] %in% neg_words) {0.1}
else if(kdf$word[i] %in% pos_words & n %in% neg_words)
0.5}
else {0.5}
#keyword count perc
#kpc <- round((nd@data$count/sk)*100, 2)
```

```
r <- rbind(r, x)
} #sub-loop ends
kdf$rank <- r
colnames(kdf) <- c("node", "word", "count", "rank")
#sums of ranked keywords
ky1 <- sum(kdf$count[kdf$rank == 1])
ky2 <- sum(kdf$count[kdf$rank == 0.5])
ky3 <- sum(kdf$count[kdf$rank == 0.1])
#weights are calculated by 'tempering' the high scores
# high score are adjusted by dividing them by sum of
remaining scores
# then multiplying that total by high score
wt <- as.numeric(ky1/(ky2+ky3))*ky1
k <- data.frame(as.character(nd@data$node[1]),as.
numeric(wt))
node_score <- rbind(node_score, k)
#remove any vertices with fewer than 4 keywords
if(nrow(kdf) >= 4) {node_list[[i]] <- kdf}
} #loop ends
#remove any empty list items
> node_list <- node_list[node_list != "NULL"]
> colnames(node_score) <- c("node", "sum_q5_kywds")
```

Now we can look at the node_score list, including only those non-empty list items:

```
> node_score[!is.na(node_score$sum_q5_kywds),]
```

Incrementing the node ranks

Now we can iterate over the vertices that do not have keywords attached to apply a score based on the size of the network clusters into which they have been grouped. Where a vertex with a keyword weighting is present in that cluster, the non-scored vertices also receive a share of the keyword score. This means that all vertices receive a score by which they can be ranked; those that are part of dense clusters that contain one or more high-ranked vertices each benefit maximally from their proximity.

This approach to shared vertex weighting constitutes a 'design proxy' in modelling community network attraction. The principle for this proxy is that a location (represented by a vertex) that people find highly attractive, such as a safe place for community life, has a positive impact on nearby locations. We could model a weighting smoothing by degree, whereby locations receive less of the weighting share based on the step they are away from the high-rank vertex. However, this would impose a computational cost to the weighting function, and possibly render the design proxy difficult to trace for observation. Instead, a proxy based on *equal* share of weighting within a vertex cluster is more efficient to compute and perhaps easier to trace.

To construct the ranking function, create an empty data frame to contain the ranks:

```
> node_rank <- data.frame()
```

Then write a 'for loop' to iterate over each of the vertex clusters, and append to them the basic network statistics that we need to calculate the rank:

```
> for(i in 1:length(clst)) {
clst_df <- data.frame(clst[i])
colnames(clst_df) <- "node"
#attach means scores to cluster nodes
clst_df$mean_score <- node_score$sum_q5_
kywds[match(clst_df$node, node_score$node)]
clst_df$mean_score[is.na(clst_df$mean_score)] <- 0
#sum of non-NA weights
s <- sum(clst_df$mean_score)
#node stats
tnc <- nrow(clst_df) #total nodes in cluster
tnn <- nrow(meta) #total nodes in network
nwn <- nrow(clst_df[clst_df$mean_score == 0,]) #total non-
weighed nodes
#wn <- tnc-nwn #total weighted nodes
#nodes in cluster as % of all network nodes
pc <- round(tnc/tnn*100, 4)
#default weight
clst_df$node_weight <- clst_df$mean_score
#adjust weights to non-weighted nodes
# nodes in clusters with NO weighted nodes each get
value:
# even share of cluster vertex count as % of all
network vertices
# non-weighted nodes in clusters with ANY weighted
nodes get value:
# an even share of cluster plus share of sum
highest-scored keywords
if (s == 0) {clst_df[clst_df$mean_score <= 1,]$node_
weight <- pc/tnc}
if (s > 0) {clst_df[clst_df$node_weight <= 1,]$node_
weight <- pc/tnc + s/nwn}
#bind to data frame
node_rank <- rbind(node_rank, clst_df)
} #loop ends
node_rank$mean_score <- NULL
```

Table 6.2 List of weights applied to the vertices

node	*node_weight*
PP 26	3.951231
bc6	3.184211
bl5	3.951231
j45	3.951231
j42	3.951231
j44	3.951231
j64	3.951231
crl6	3.951231
bjl2	3.951231
j46	3.951231
ur32	3.951231
sp28	19.012500
shch29	3.951231
na69	1.428575
j60	1.428575
j36	1.428575
j57	1.428575
j56	1.428575
j47	1.428575
ap0	1.428575

The approach to ranking vertices based on keyword scores, or else on proximity to weighted vertices within vertex clusters, described in this chapter results in a vector of vertex weightings. Table 6.2 provides the first 20 nodes and the ranking scores from our current sample.

Constructing the directed graph

The vector of ranking scores allows us to automatically generate a directed graph of community relationships with local spatial assets. The direction of network edges represents how communities are attracted to or repelled from certain locations due to the influence of positive or negative connotations, respectively. Constructing a directed graph involves choosing which vertices the direction is moving 'from' and which it is moving 'to'. Hence, the direction will move from the lower-ranked vertices and to the higher-ranked vertices. This principle for direction stands as a proxy for where community members are likely to move towards, perhaps given special circumstances for movements. An example of this might be a disaster or emergency, in which people move towards places they deem to be safe.

Firstly, we create a new look-up data frame to contain the ranking scores we need to plot the directed graph:

```
lkp_df1 <- merge(lkp_df, node_rank, all=T)
lkp_df1$mean_score <- NULL #no longer needed
lyt1 <- as.matrix(lkp_df1[,2:3])
```

Then we create an empty data frame that will contain the vertex pairs for the directed graph. A 'for loop' will iterate over the vertex pairs, record the rank and assign the vertex with lower rank to a 'from' vertex of the directed graph, and the higher to a 'to' vertex. These 'from-to' pairs are then bound to the data frame:

```
from_to <- data.frame()
for(i in 1:nrow(node_pairs)) {
v1 <- lkp_df1$node_weight[match(node_pairs[i,]$node1,
lkp_df1$node)]
v2 <- lkp_df1$node_weight[match(node_pairs[i,]$node2,
lkp_df1$node)]
from <- if(v1 <= v2) {node_pairs[i,]$node1} else
{node_pairs[i,]$node2}
to <- if(v1 > v2) {node_pairs[i,]$node1} else {node_
pairs[i,]$node2}
ft <- data.frame(from, to)
from_to <- rbind(from_to, ft)
}
```

We can now build a graph data frame based on the from-to pairs, using the new look-up data frame for plotting. A simplify function will also remove any network loops that may have been introduced. This can then be plotted along with a street-network data sample:

```
digraph <- graph.data.frame(from_to, vertices = lkp_
df1, directed = T)
digraph <- simplify(digraph)
```

This geo-located directed graph can also be plotted with additional geographic data, as in Figure 6.4, that includes a street-network sample. In this example, we can see how the local street network bounds two open spaces (which are both public parks). One of these to the right of the plot was selected by many participants in the study for largely positive connotations. Another nearby public park was also selected by many participants but for largely negative connotations. This is reflected in the directions of the network edges, showing how one park seems to draw local community members, and the other pushes people away in some, but not all, aspects.

Conclusion

In this chapter we have outlined a method for describing community relationships – including those between people, places and artefacts – in the form of a network graph. We developed a method to generate an unweighted, undirected network graph. We then applied weights to the graph vertices stemming from community descriptions of their associated spatial assets. We developed a method

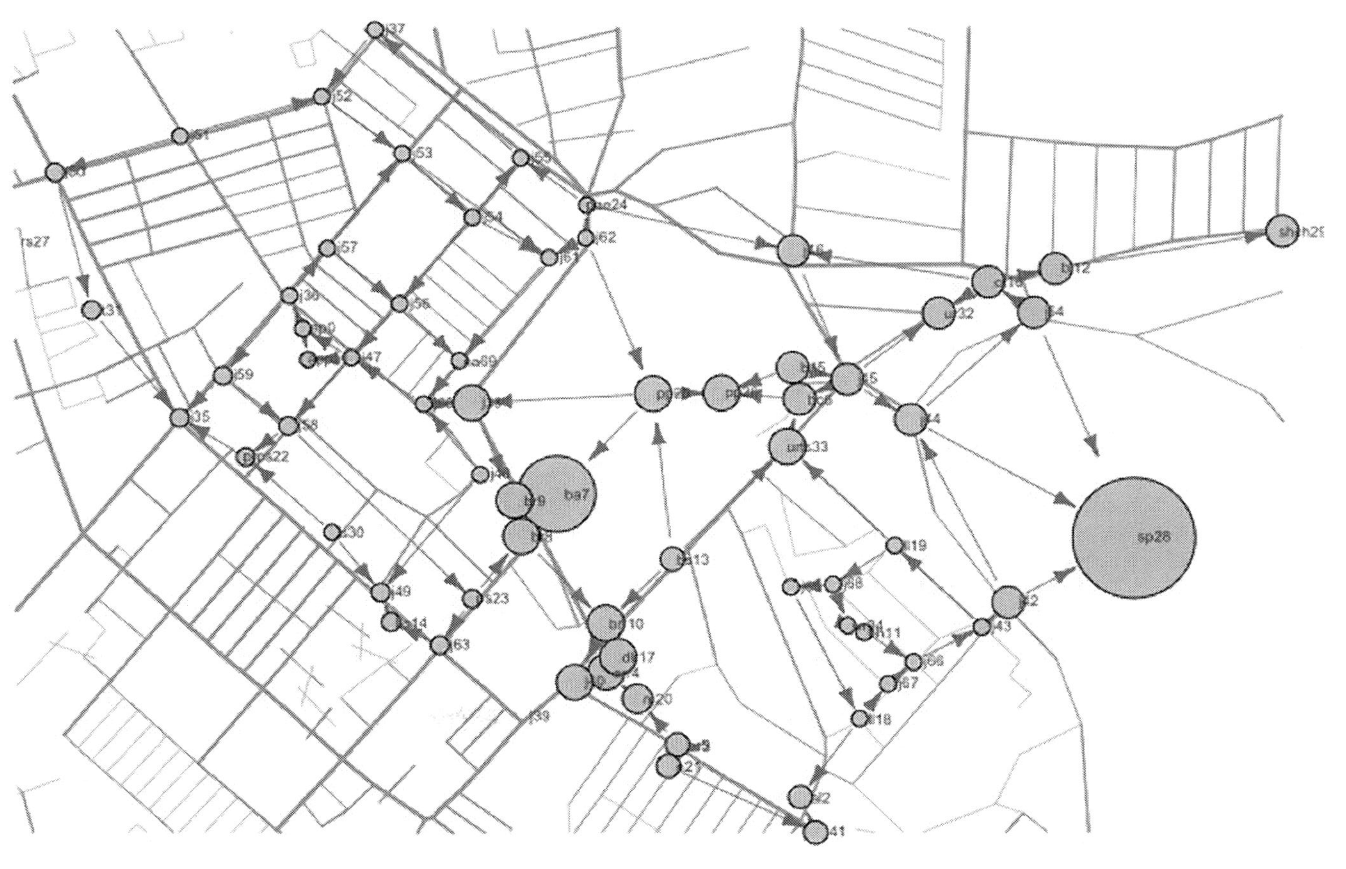

Figure 6.4 Directed graph of relationships among selected community assets, vertex sizes adjusted for community weightings.

for these weights to be shared among neighbouring vertices, which could benefit from their network proximities. A further development of this method could apply an inverse weighting metric, so that vertices receive higher or lower weightings the closer or further they are away from the weighted vertex.

We have outlined how this toolkit can be developed using the R programming language and development environment, including the iGraph library. Any similar programming environment could also be used, and readers interested in developing similar network models might make use of our workflow in software development. In the next chapter, we focus on possibilities for more advanced iconographic representations of the kinds of spatial relationships described by community participants.

Note

1 Transcribing this graph can be made easier by dictating the points that connect to a colleague, who can input them directly to a spreadsheet, or else using a voice audio recording.

7 Building and representing knowledge

Urban space and knowledge representation

To introduce our thinking around the problems of knowledge representation (KR) in urban community domains, we draw out a series of arguments from a diverse range of social and spatial literature. Considered together, these arguments point to a need to 'flatten out' our 'globalized' knowledge of communities and their spaces, to reveal their material forms as well as their conceptual relationships.

Urban practitioners engage their 'design knowledge' in the problem space of the project, which has been seen traditionally as belonging to either of two frames: that of rational problem-solving and that of reflective practice (Doorst and Dijkhuis, 1995). To reiterate a theme we touched on Chapter 1, the notion of the 'rational' frame suggests that the designer forms an informational process within an objective reality and seeks optimal results from poorly structured problems. The notion of the 'reflective' frame suggests that the designer constructs the 'problem situation' through his or her creative and iterative practice. We have already observed how, design knowledge by professional and non-professional practitioners alike arguably includes both rational and reflective approaches.

Urban practitioners approach urban spaces as highly complex artefacts, which follow functional schemes shaped by the needs of movement and information at local and global scales. They are formed from 'urban images' of boundaries, thresholds and interfaces of everyday activities (Lynch, 1960; Conroy Dalton and Bafna, 2003; Palaiologou and Vaughan, 2012), from socio-spatial integration or segregation (Vaughan and Arbaci, 2011) and from ideological and political distinctions of power and control (Hillier and Hanson, 1984, p. 21). When describing the relational complexities of urban spaces, Hillier (2007, pp. 27–30) has observed differences in the modes and methods of description. Those engaging in urban forms for professional purposes often *think of* space, informed that is by urban analytical theory. Those making everyday use of urban forms *think with* space, informed by local knowledge and natural movements (Al-Sayed, 2014; Hillier, 1999).

Design knowledge requires practitioners to think of urban spaces in terms of their spatial and conceptual associations or implications, or rather their 'spatiality' and 'trans-spatiality' (Hillier and Hanson, 1984, pp. 40–41; Sailer and Penn, 2009). Practitioners categorize urban spaces by combining representations of *cognitive and historical experiences* of artefacts through their associations and implications (cf. Lefebvre, 1991, pp. 294–297), within the limits of spatial and temporal logics (ibid, pp. 195–196). From these experiences and logics they extend categorical 'image schemata' to form representations of their spatial *cognitive and historical knowledge* (MacEachren, 2004, pp. 185–190).

To support our practical thinking, we make use of visual and spatial metaphors such as circles, triangles, planes, globes and scales. However, these metaphors can serve to enframe our thinking, possibly based on the particular viewpoint of the dominant group, for example industrialized, rational, male, literate and so on (cf. Ingold, 2000, pp. 209–218). Academic discourse in this field has been criticized for hiding its knowledge behind such 'frames' (Hommels, 2010), and arguments have been made to unfold or flatten out knowledge (cf. Ingold, 2000, pp. 189–208, 2011b, pp. 229–243), to represent an urban environment with all elements unhidden and intelligible (Hillier, 2007, pp. 67–68), and without the impositions of abstract schemata (Lefebvre, 1991, pp. 301–302).

Representing domain knowledge

In this section we explore the possibility that concept graphs provide a method of representing spatial and trans-spatial knowledge. Working with concept graphs allows us to model semantically rich domains that can include sets of beliefs, desires and intentions among community participants (Sowa, 2008; Kavouras and Kokla, 2007). Logic can be used to model diverse community formations in a consistent and dynamic way.[1]

Concept graphs are based on simple logical constructions of domain knowledge, providing an intuitive and portable schema for KR. Graphs as logic tools have been described as 'building blocks' for expressing knowledge in terms of entities. They provided a set of tools for testing (for 'true' or 'false') entities – whether facts, goals, implicit or explicit rules – their attributes, conditions and relationships (Chein and Mugnier, 2009, p. 22).

A set of standard statistical methods in graph analysis allows us to analyze relationships among entities based on conceptual distances, measured by degrees, densities and centralities. Similarly, a set of established layouts allow us to represent these relationships based intuitively on patterns of flow, force, orientation and geometry, as well as colouring and labelling. Employing these methods may reveal the functional significance of an entity or cluster of entities within the network model, which may also reveal the significance of conceptual entities whose roles may not be readily apparent from real-world domains.

Concept graphs can represent entities and relationships using generalized or categorical concept nodes and relation nodes. Every concept has an abstracted type, which can be either specified or non-specified. Concept graphs can be used to configure and test assertions by 'projecting' or 'simplifying' sets of abstract concepts into specific instances (and vice versa).

Conceptual graphs represent this schematization using arcs (or edges) that connect concepts to relations:

```
[Concept_1] -> (relation) -> [Concept_2]
```

Concept and relation types are arranged hierarchically based on a generalization order, meaning that one type can subsume another. For example, *girl* (type) would bear the same characteristics as *person* (type); in other words, *person* subsumes *girl*. The universal type is marked by the sign T:

```
Alice < Girl < Person < Human < Organism < Entity T
Sleeps < Resting activity < Activity < Action < Event T
```

Multiple type structures can support semantic conjunctions (for example, 'the girl Alice sleeps in the park which is sunny and pleasant'):

```
[Girl:Alice] -> (Sleep) -> [Park:Sunny:Pleasant]
```

Graphs can be projected in the sense that their nodes can be changed into specific sub-types or general super-types and then tested for logical composition. Projection also supports graph unification, whereby sets of nodes between graphs are generalized (preserving their arguments and values), and then compared with similar graphs to identify their similarities (isomorphisms).

The example conceptual graph in so-called linear form (below) demonstrates how the generalized graph projects into the specialized graphs. This shows how agents interact with entities (and their themes), and also their position within the conceptual hierarchy: {*} demarcates the top supertype and *x the bottom subtype (being the specific instance; sometimes called the 'absurdity'), while ?x can demarcate uncertainty:

Generalization:

```
[Person:Name {*}] <- (Agnt) -> [Activity:Resting] ->
(Thme) -> [Space:Weather:Feeling]
```

Projections:

```
[Girl:Alice *x] <- (Agnt) -> [Sleep *x] -> (Thme)->
[Park:Sunny:Pleasant *x]
[Boy:Billy *x] <- (Agnt) -> [Shelter ?x] -> (Thme) ->
[Park:Raining:Cold *x]
```

Using these and other logic tools, conceptual graphs allow high-level generalizations to be agreed among a community of domain practitioners, and specialized with more specific or concrete instances of those general categories. Conceptual graphs may be constructed through a top-down (general-to-specific) or bottom-up (specific-to-general) process. Importantly, a 'middle-out' extension to this well-established graph process has been developed by Berta et al. (2016) in the field of 'urban ontologies'. Here (to paraphrase), the specifications of relational concepts (ontologies) are extrapolated through domain practices. The extrapolation process is limited selectively according to the *scale of representation*, the *historical significance* and the '*relational functionality*' (in terms of logical composition) of the urban elements under analysis. The ontologies are shared as a compositional template, which includes spatial and functional classes, spatial data properties and their logical relationships from a given set of domain phenomena.

Representing spatial logics

Spatial logics of urban configurations have been described in various urban-domain contexts, in terms of both Euclidean (abstract) and topological space. For example, urban spaces have been shown to bear structural and geometrical patterning based on metric and topological distances and local-global relationships (Hillier, 1999; Hillier et al., 2010; Hillier and Vaughan, 2007). In another field, the region connection calculus is based on a set of eight basic (abstract) relations of connection, intersection and contact (Randell, Cui and Cohn, 1992). These can be used to describe part-component relationships (A and B is *part-of* C, or C is *made-of A and B*), including meronymic relationships[2] where entities coincide within their parent class (Berta et al., 2016).

The language RCC8 is a formalism to describe eight basic spatial relations, which can be rendered visually, as shown in Figure 7.1.

We may also offer icons for Lakoff's (1987) image schemata of spatial relationships that are not available in RCC8, as shown in Figure 7.2.

Proof-of-concept

A research exercise was undertaken to test the practicable viability for urban-domain KR of incorporating spatial logics using RCC8 into a concept graph

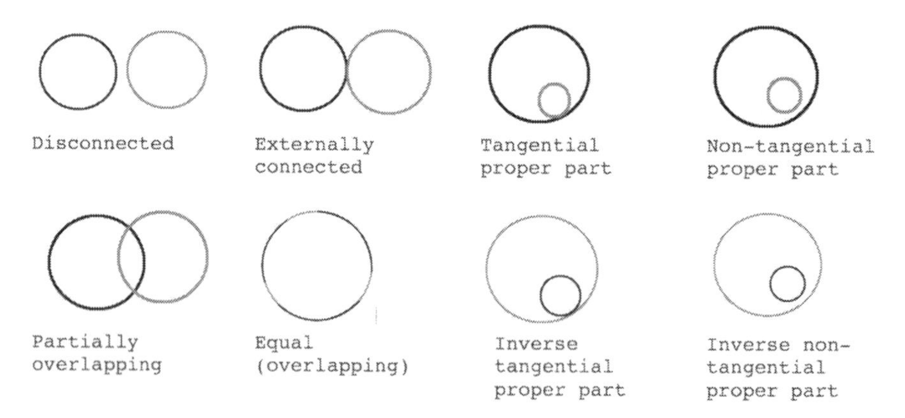

Figure 7.1 Visual representations of the RCC8 language for spatial relationships.

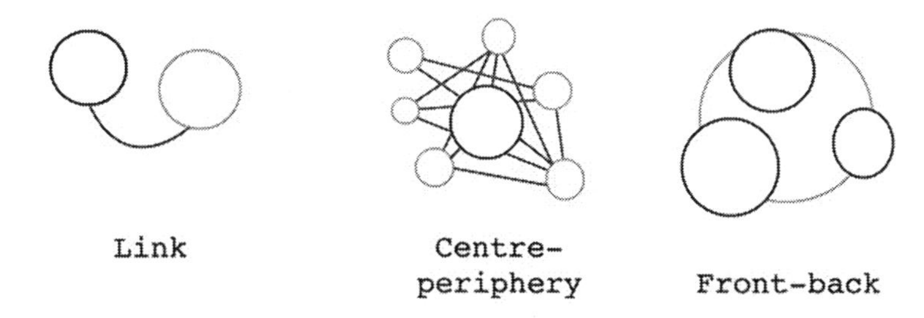

Figure 7.2 Visual representations derived from Lakoff's schema for describing spatial relationships.

schema. Towards this aim, we conducted a prototyping workshop involving a small group of planning and design practitioners.[3] They were invited to engage with mapped data visualizations from an earlier data-gathering exercise with secondary school-age children across Liverpool (O'Brien et al., 2016), which was part of the broader study at University College London, 'Visualizing Community Inequalities', already outlined in Chapter 3. The participating children had 'mapped' their local urban communities by selecting significant features and weighting these by applying emoticon stickers to local maps. These maps and emoticons were digitized with a GIS and the points data were manipulated to produce the visualizations.

The urban practitioners were then presented with a set of graphic icons that represented the range of structures, scales and other features selected by the school participants. The participants were also presented with graphic representations of spatial logics (as above) to describe ways in which these structures and scales might be arranged. The participants were then invited to build a basic (roughly defined) concept graph to describe any discernable patterns of community formations within the mapped data. Here we present just one example of the concept graphs produced by the participants.

Figure 7.3 shows one example of ways in which the participants configured conceptual graphs to schematize the participatory spatial descriptions. The icons have been arranged to represent a journey from home (left) to a supermarket (right), which crosses busy roads carrying local (pedestrian/velomobile), city-wide (light/heavy automative) and regional (heavy automative/transit) traffic. The journey has a negative dimension involving a road junction (for example, for hindering pedestrian access). The supermarket contains a café as a positive dimension (for example, for social life), represented here using an RCC8 icon for 'tangential proper part' (TPP). From this illustration we can construct a generalized and specified concept graph:

Generalization:

```
[Origin:Name] <- (Agnts) -> [Activity] -> (Thme) ->
[Urban entity:Type] -> [Movement:Scales] -> (Thme) ->
[Urban entity:Type] -> [Logic relation:RCC] ->
[Destination:name]
```

Figure 7.3 This participant's concept graph represents a possible journey from home to a supermarket.

Projections:

```
[Home-shop] <- (Parent+child) -> [Cross road] -> (Neg) ->
[Junction] -> [Traffic:Local+Citywide+Regional] ->
(Pos) ->
[Shop] -> [RCC8:TPP] -> [Café]
```

Other participants in the workshop were able to use RCC8 icons to describe urban community relationships in terms of being 'externally connected', 'disconnected' and 'tangential proper part' and 'non-tangential proper part' (NNTPP). Interestingly, the latter instance of NNTPP was used to refer to an activity taking place in a public park, which perhaps speaks to the production of a local social space (a sporting activity) that is part of, but not physically integrated with, the public open space. From this and other examples of successful graph constructions, there remains a possibility for developing and refining this prototyped technique.

Conclusion

The proof-of-concept exercise described in this chapter demonstrated the viability of incorporating spatial logic schema in a concept graph. Participants' engagement with a set of graphic icons (representing urban spatial entities, movement scales and spatial relationships) provided an intuitive method for representing their observations of complex data. Towards the objective of creating a practicable tool for KR in urban domain practices, we next need to test how the graphic icons can be organized into a meaningful 'flow' to support domain diagnostics and decision-making.

One possibility in this area is to arrange these icons within an argumentation schema. Arguments often derive from expert opinion, from metaphors, analogies or precedents or from practical reasoning (and, negatively, from ignorance, misinformation or prejudice). Challenges to an argument can be made by posing critical questions that serve to interrogate the argument's assumptions, premises and logical formulations (Walton, 2013, p. 28). The field of argumentation provides a range of informal logic schema for enriching and testing representations of domain arguments.

Notes

1 A lucid introduction to concept graphs has been provided by Polovina, 2007.
2 'Meronymic' refers to parts being joined to the whole via a structurally functional connector (Berta et al., 2016).
3 An illustrated description of the workshop is available at http://tinyurl.com/hqkv7jv. The author is grateful to staff at URBED Ltd, Manchester (UK), for supporting this workshop.

8 Cases in urban community formations

In this chapter, we outline ways in which bringing several different perspectives on urban communities might help to illuminate the social and spatial contexts for their formations. By presenting representations of community contexts as maps and diagrams, we intend to shape themes in design research in the area of urban community formations. Some methods based on community networks are experimental, and further refinement may help hone their usefulness as design-thinking tools.

The perspectives are represented throughout this chapter as maps and network diagrams. In Chapter 3 we outlined a study undertaken at University College London into urban community formations among secondary school-age children in Liverpool, UK. This study resulted in a series of data sets pertaining to the community members' perspectives on their local spaces. In this chapter, we highlight two of these data sets that reveal perspectives on community spaces in contrasting areas of the broader urban context.

These contexts are described in terms of socio-economic and geographic status by statistical ranking, gross patterns of land use, some of the participants' perspectives on their community spaces and directed and undirected graph models of community formations. We also focus on two types of spatial assets that are deemed to be most significant by the participants in each case study, namely roads or junctions and open spaces, being public parks in the majority of instances.

The community case studies

North Group and South Group are each formed from pupils at two schools in, respectively, northern and southern areas at the edges of the urban centre of the city of Liverpool. Figure 8.1 shows their locations with reference to the (sampled) Merseyside conurbation. The space between the major urban centres shown in this figure represents the River Mersey. The two spanning network structures represent tunnel complexes for vehicle traffic. These groups were not identified for the purposes of the research study as being especially discrete or even well defined by the study participants. Yet the study revealed how the groups of 11- to 18-year-olds that made up the participant groups habitually made use of common spaces, and they shared many perspectives, perceptions and experiences. They

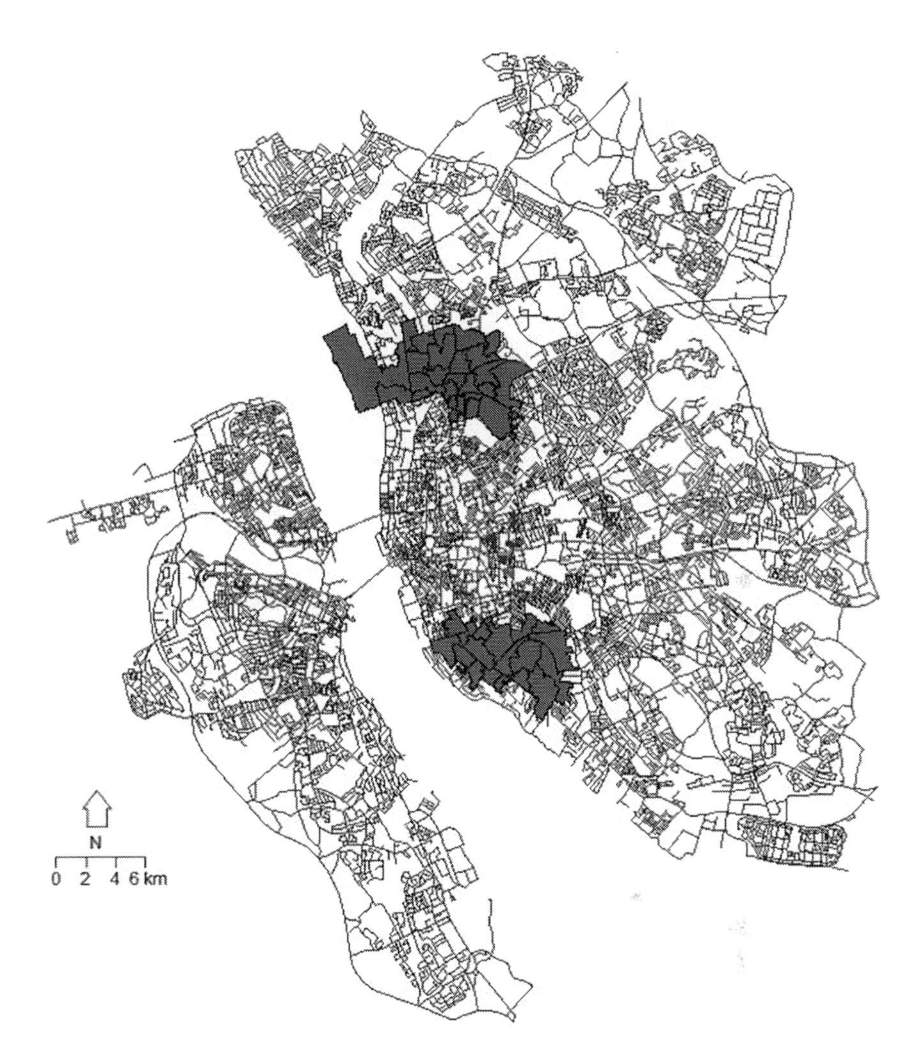

Figure 8.1 The LSOAs for the case-study sites of North Group and South Group in the broader urban context of the sampled Merseyside conurbation.

also experienced conflicts within these spaces, which is perhaps a natural component of community life in any context.

The contexts for North Group and South Group community formation contrast in several aspects. Firstly, we can take a broad look at the socio-economic profiles of the case-study contexts as measured by the Index of Multiple Deprivations (IMD), which is based on UK census data from 2015 (Office of National Statistics). The IMD combines several indicators of deprivation, including income, unemployment, quality of the environment, chronic illness and so on. As such, there is only an indirect relationship between the IMD rankings of a particular area and the experiences of the people who live, work, go to school or make use of local assets in that area. However, the IMD can provide a useful indicator for the relative paucity of community assets in a given area.

We can see from the plots in Figure 8.2 and Figure 8.3 how the sampled communities are situated in areas presenting overall moderate to high deprivations. This does not mean that the participants themselves directly experience these levels of deprivation (many of the school children live outside of these areas), though undoubtedly some of them will experience deprivations. Our interest in plotting the community spatial context in this way is to infer, or improve our understanding of, the relative attractions and affordance for community resources in these areas.

The two community contexts also compare in the spatial patterning of deprivation distributions. Each sampled area presents hard edges between IMD presentations in different deciles. The deciles represent very wide IMD scores, so there can be a stark difference between an area in the first (bottom) and in the second decile. Those living in areas of the first decile are likely to experience severe hardship and possibly of social marginalization. This is not to say that people in these areas do not enjoy community life, or make use of community assets, including places of worship. However, they are less likely to live close to well-resourced, essential assets. These can of course affect anyone whose community life is situated in this area, even if they are not ordinarily resident in the area.

For this reason, a useful sub-domain of IMD relates to the measurement of geographic barriers, as defined by the UK Office for National Statistics. This index is extrapolated from a set of spatial metrics in each area sampled, including:

- Road distance to a post office
- Road distance to a primary school
- Road distance to a general store or supermarket
- Road distance to a doctor's surgery

Figure 8.4 and Figure 8.5 show these indices in terms of their national decile positions, where 1 represents the decile with the highest geographic barriers (bottom decile) and 10 the lowest (top decile). The decile values are represented in dark-to-light shades, with black representing decile 1, or the highest geographic barriers. The community sites, where the research workshops were facilitated, are roughly in the centre of each sampled area. In each example, the community's spatial centre is located in an area presenting high geographic barriers, involving greater distances to the key assets listed above. Overall, these areas present generally low to moderate geographic barriers, which perhaps relates to their proximities to high-integration centralities.

Land-use samples allow us to look more closely at the nature of the geographic barriers in the North and South Group contexts (Figure 8.6 and Figure 8.7). In both cases, parkland presents the main geographic barrier (shown in diagonal lines within polygons). In the North Group, two railway branch lines (shown as thick white lines) bound and bisect the community space. The spatial edges

Figure 8.2 Index of Multiple Deprivations (IMD) for North Group spatial context (LSOA by decile, ONS 2015), where 1 is the most deprived.

Figure 8.3 IMD for South Group spatial context, as above.

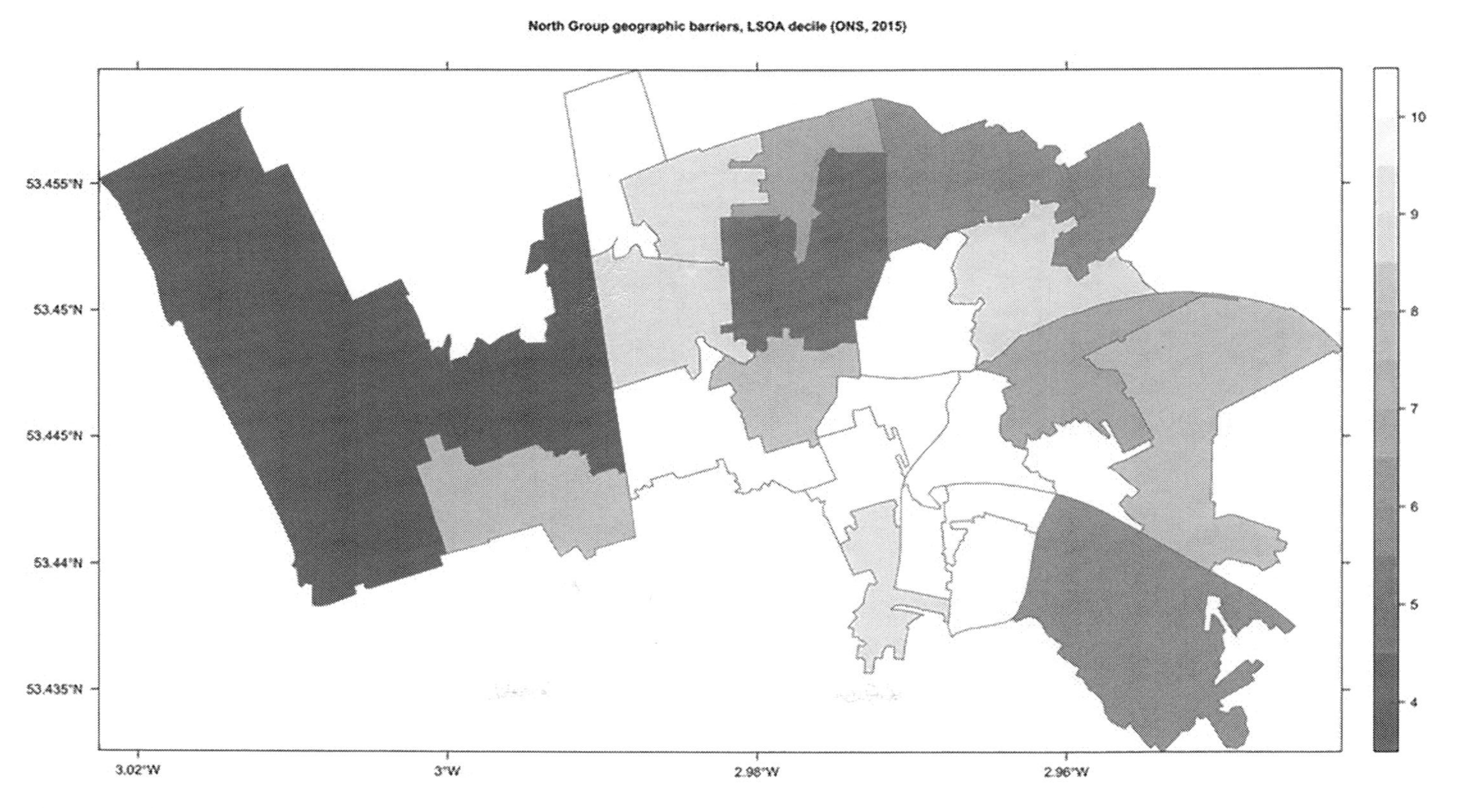

Figure 8.4 Geographic barriers for LSOAs within the North Group community area.

Figure 8.5 Geographic barriers for LSOAs within the South Group community area.

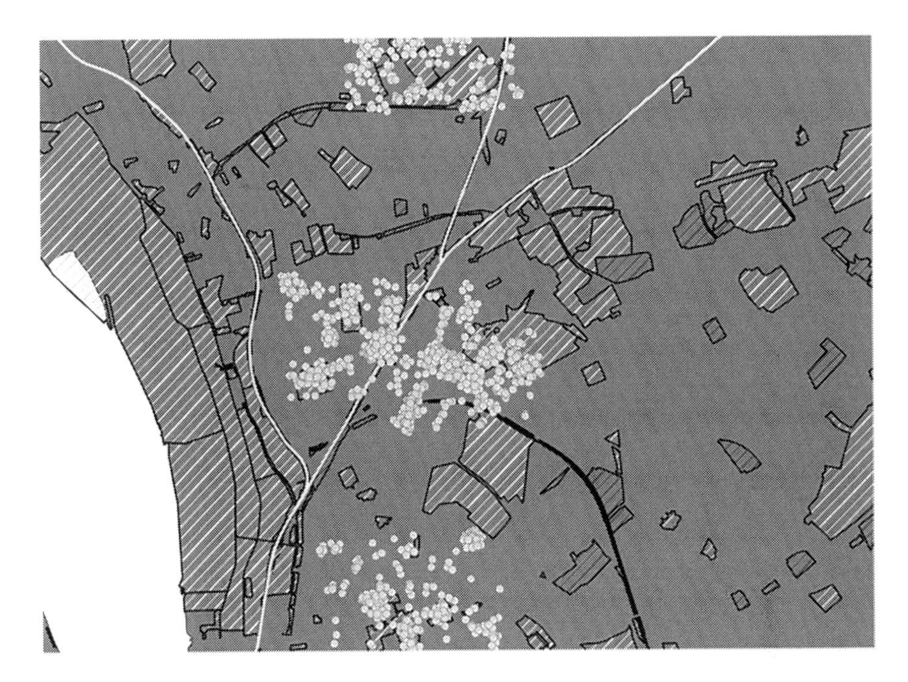

Figure 8.6 Schematic map of land use (all use types except residential) for the North Group community area.

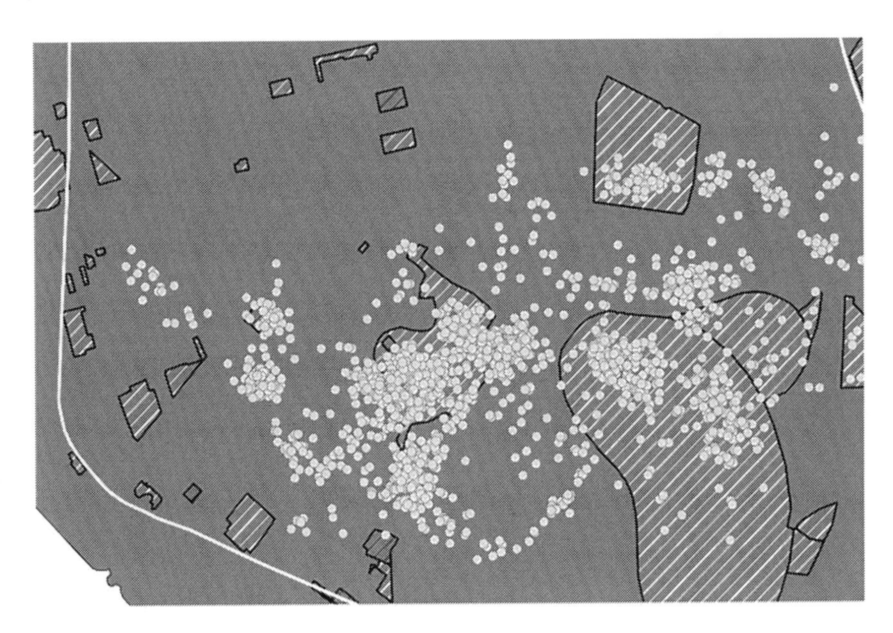

Figure 8.7 Schematic map of land use for the South Group community area (as in Figure 8.6).

formed around these railway lines are hardened by strips of inaccessible scrub. A large industrial zone forming part of a riverside dock complex also presents a hard edge to the west. Parkland in the South Group context presents the major geographic barrier to accessibility, with a several public parks forming barriers to the wider local area. Contrasting the geographic contexts of the North and South groups, it is evident that the North Group's spatial context presents a set of hard barriers to spatial accessibility.

Road-network structures

Figure 8.8 and Figure 8.9 show the major features of the community spaces selected by the participants, represented as the centroids of the clusters of digitalized points. These have been plotted over a road-network map, which represents the range of network integration, where thicker lines represent higher integration. We can regard this map as showing how well inter-connected these network segments (centralities) are to the wider road network. This also suggests how much traffic is likely to occur along these centralities (for example, intersections of thicker lines represent a high likelihood of heavy traffic occurring around these junctions).

The selections made by the specific sub-groups that constituted the participant communities (that is, the schools where the workshops were held) are represented by white and black dots, respectively. Figure 8.8 shows how the North Group falls distinctively into two sub-groups, with some convergence in their focus around busy road junctions. Community formations seem within road sub-networks, and are bisected along a high-integration segment (the location of a busy main road).

Figure 8.8 Road-network integration for the South Group community area with major features selected by the two community sub-groups.

Figure 8.9 Road-network integration for the South Group (as in Figure 8.8).

The darker network segments represent high integration of network centralities within a 2km radius. The major features selected by the two community sub-groups are represented in black and white dots, respectively.

The South Group plot of points centroids shows a greater degree of convergence in focus among the two sub-groups. In Figure 8.9 we can see how the recti-linear layout pattern serves to form a boundary for community formations in this context, appearing to push the community into a discrete space, focused around a major junction, a road that forms an edge between two public parks and around the parks themselves.

Observing the distributions of points centroids over the integration network provides initial 'snapshots' of community focuses in the context of the research. We can begin to get a sense of the attractions around roads, junctions and open spaces. However, to gain a keener sense of community formations among these groups, we need to derive greater semantic richness from the available data.

Semantic associations

We can firstly subset the data based on 'negative' and positive' associations. The participants used a range of emoticon icons to code their selections, which included symbols to represent sentiments associated with happiness, love, sadness and anger, as well as perceptions of safe places or hazards, and so on. These were used to represent a range of basic emotions, which were derived from a standardized psychological schema and perceptions of hazards, risks, barriers and so on. A basic aggregation of these data, based on these semantic associations, reveals a mixture of positive and negative association for many, or even most, of the features selected. In both North Group (Figure 8.10) and South Group (Figure 8.11) there appears to be a higher prevalence of negative associations over positive associations, perhaps because feelings of anger, fear, sadness and so on are more

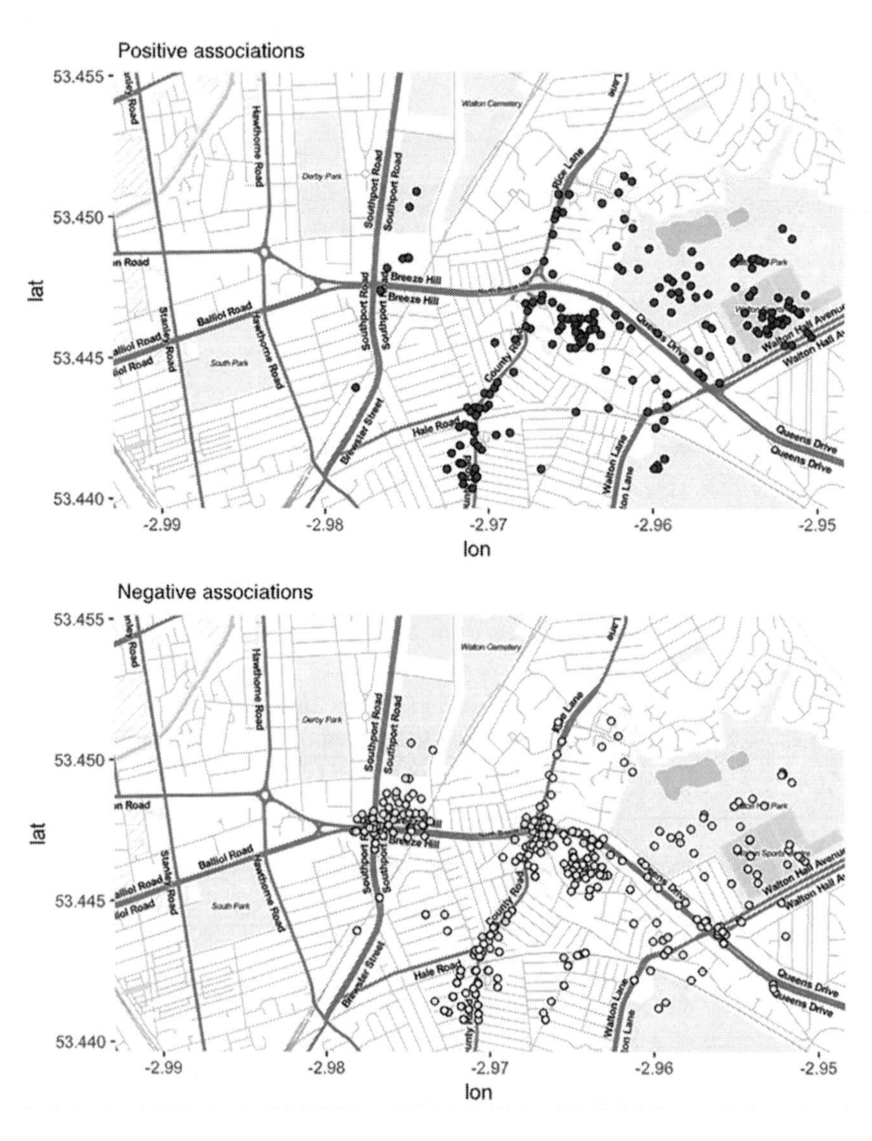

Figure 8.10 Distributions of points aggregated by their 'positive' and 'negative' associations, respectively, for the North Group.

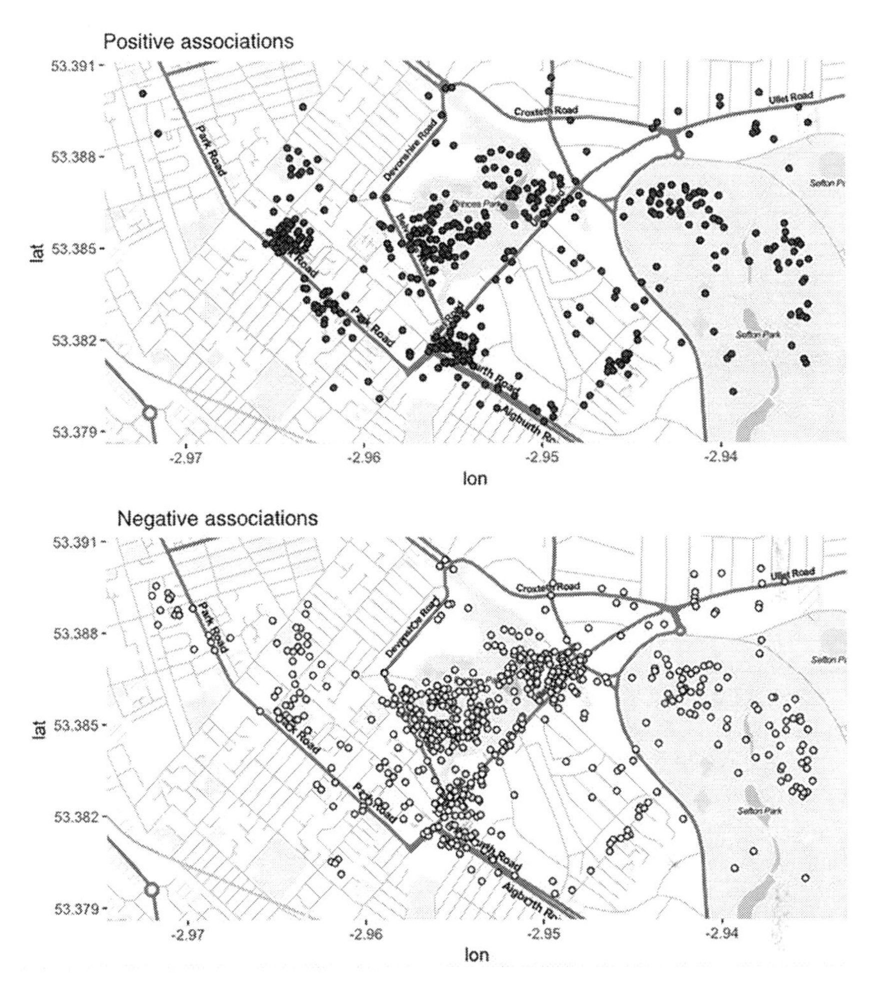

Figure 8.11 Distributions of points aggregated by their 'positive' and 'negative' associations, respectively, for the South Group.

readily available for articulation in the context of the built environment than those of happiness, love, safety and so on. We can also observe that in both community contexts, there is a high prevalence of negative associations relating to high-integration road segments (probably very busy or congested) road junctions.

The next step in our analysis is to focus on the major features selected by the participants, including how these differ among the age groups. We can achieve this by arranging the data into a series of tables (Tables 8.1a and 8.1b, and Tables 8.2a and 8.2b). These show how much 'weight' has been applied to these features, or how many emoticons have been applied to the map location of this feature by the group as a whole.

Table 8.1 (a and b) Tables of the frequencies of main features selected by the North Group community participants, arranged by age group

(a) **12–13-year-olds**

Type	Frequency	% total
open space	108	12.69
school	104	12.22
junction	43	5.05
leisure	43	5.05
road	42	4.94
shops	31	3.64
area	14	1.65
residence	11	1.29

(b) **14–15-year-olds**

Type	Frequency	% total
road	136	15.98
school	79	9.28
open space	69	8.11
junction	35	4.11
leisure	21	2.47
area	14	1.65
roundabout	13	1.53
residence	9	1.06

For North Group, many of the 12- to 13-year-olds selected features under the category of 'open space', which were mainly public parks. Other major features among this group included schools, roads and junctions, leisure facilities including sports centres, and shops. These types of features were also prevalent among the 14- to 15-year-old sub-group, but open spaces were selected on few occasions compared to the younger group.

We could possibly remove 'school' from the analysis of the community's urban context, as this constitutes the internal space for the community, with a set of spatial and social dynamics that are independent of the wider local community spaces. Once 'school' has been removed from the list of types, these tables serve to make clear that roads and junctions (including crossings and roundabouts) and open spaces are the predominant features for community formations among the North Group participants.

A very similar pattern of selection is in evidence among the South Group participants. Open spaces are the predominant features selected by participants in the three age groups. Schools are also major features, along with roads and junctions, and also a local café/restaurant. This community meeting place is important to varying degrees among all age groups. Given the similarity of degrees in prevalence for open spaces, roads and junctions among all participants, this can be the focus of our next stage of analysis.

Sentiment mapping of community formation

In the previous section we noted how open spaces form the predominant type of feature selected by the workshop participants. We can now focus on some of the semantic dimensions of this points data subset. We have already outlined how participants in the study made use of a selection of emoticon stickers. This yielded a highly rich and diverse data set, from which we inferred some implicit patterns in community formation within the case-study contexts.

Table 8.2 (a–c) Tables of the frequencies of main features selected by the South Group community participants, arranged by age group

12–13-year-olds

Type	Frequency	% total
open space	140	13.13
school	100	9.38
road	86	8.07
shops	46	4.32
transport	31	2.91
junction	25	2.35
café/restaurant	21	1.97
crossing	20	1.88
area	17	1.59

14–15-year-olds

Type	Frequency	% total
open space	92	8.63
school	53	4.97
road	32	3.00
shops	32	3.00
junction	22	2.06

16–17-year-olds

Type	Frequency	% total
open space	106	9.94
road	44	4.13
school	41	3.85
junction	34	3.19
café/restaurant	15	1.41
place worship	14	1.31

Where emoticons related to open spaces (Figure 8.12 and Figure 8.13), in each case these were specific public parks close to the schools where the workshops were conducted. Coincidentally, each community made selections of two parks, and each yielded more negative associations than positive. This is perhaps surprising, as designers and planners might often consider green spaces in urban contexts as offering a given benefit to local communities. The participants in the workshop studies presented some text descriptions of their selected local assets, which may provide evidence for the nature of the negative and positive associations.

It is also worth observing how three of the four parks selected by the sampled communities are edged by main roads. The road-network analysis using space syntax that we outlined in Chapter 4 revealed how these structures present high-integration values relating to radial scales of 2km to 10km. In other words, these network structures converge urban movements (vehicle traffic) from across the city and its wider region. Accessibility to these open spaces may be hampered by the proximity of these roads that, we can hypothesize, form barriers to accessing these potentially valuable assets.

Roads and movement infrastructures

Considering the potential negative impact of roads on accessibility to open spaces (Figure 8.14 and Figure 8.15), we can also observe the participants' semantic associations with these infrastructural assets. In both case studies we can observe some dominant features in the community sample, including main roads that seem to split the community spaces into discrete quarters, and major junctions that intersect these roads. We find predominantly negative associations with these infrastructural assets, and we can mine the participants' text descriptions of their sentiments relating to these.

Sentiment keywords

In order to understand what kinds of negative and positive associations the participants carry, we can simply extract keywords from their text descriptions. These can be sampled by the most widely used words and plotted as a set of histograms revealing the keywords associated with positive and negative sentiments for each community group, as we can see in Figure 8.16. These plots represent a mixture of sentiments. However, some are not entirely clear as they are separated from their semantic contexts; for example, 'happy' in the set of negative sentiments might have become separated from the phrase 'not happy'. Nevertheless, the major features of the plots make it clear how positive sentiments are closely associated with social encounters, tokenized to 'friends' or 'mates'. Interestingly, 'walk' is also associated with positive sentiments in each case-study context, reflecting the main way in which the participants get from home, or transport hubs, to their schools.

The dominance of major roads in the North Group context is also revealed in a range of positive and negative associations with these, and their impact on the urban environments. These are revealed in the tokens 'traffic', 'cross' (as in

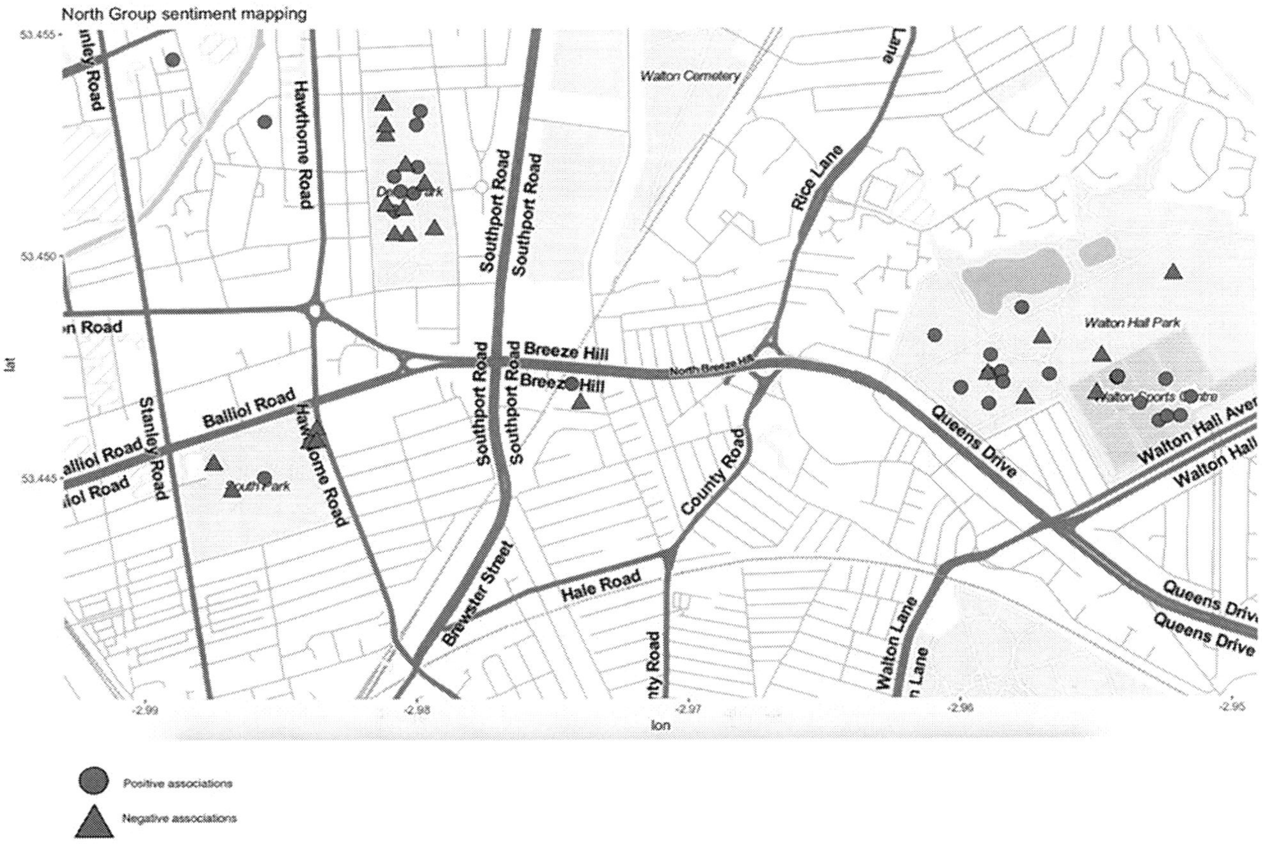

North Group sentiment mapping

Positive associations

Negative associations

Figure 8.12 Distributions of positive and negative sentiments associated with open spaces (especially public parks) for the North Group participants.

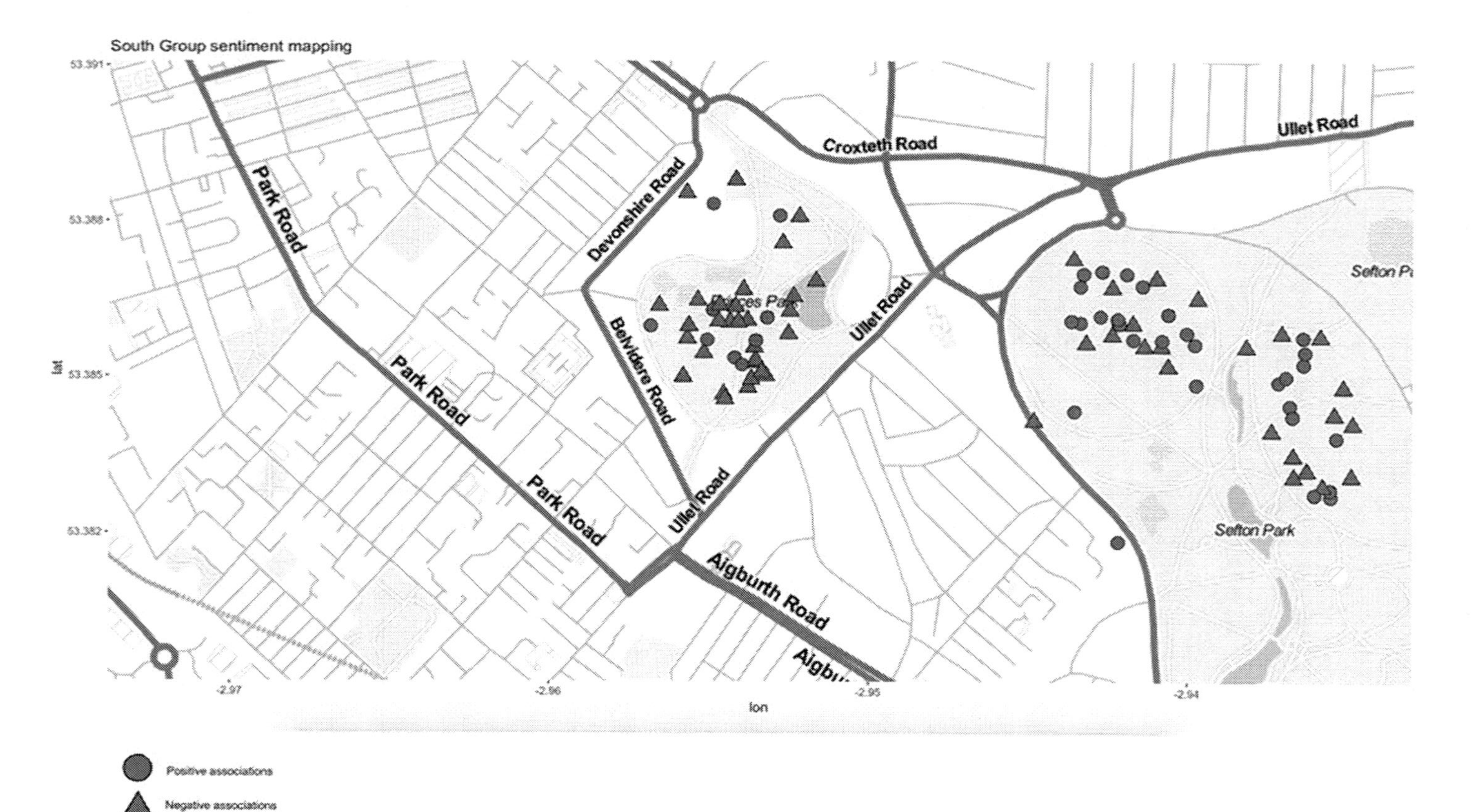

Figure 8.13 Distributions of positive and negative sentiments associated with open spaces (especially public parks) for the South Group participants.

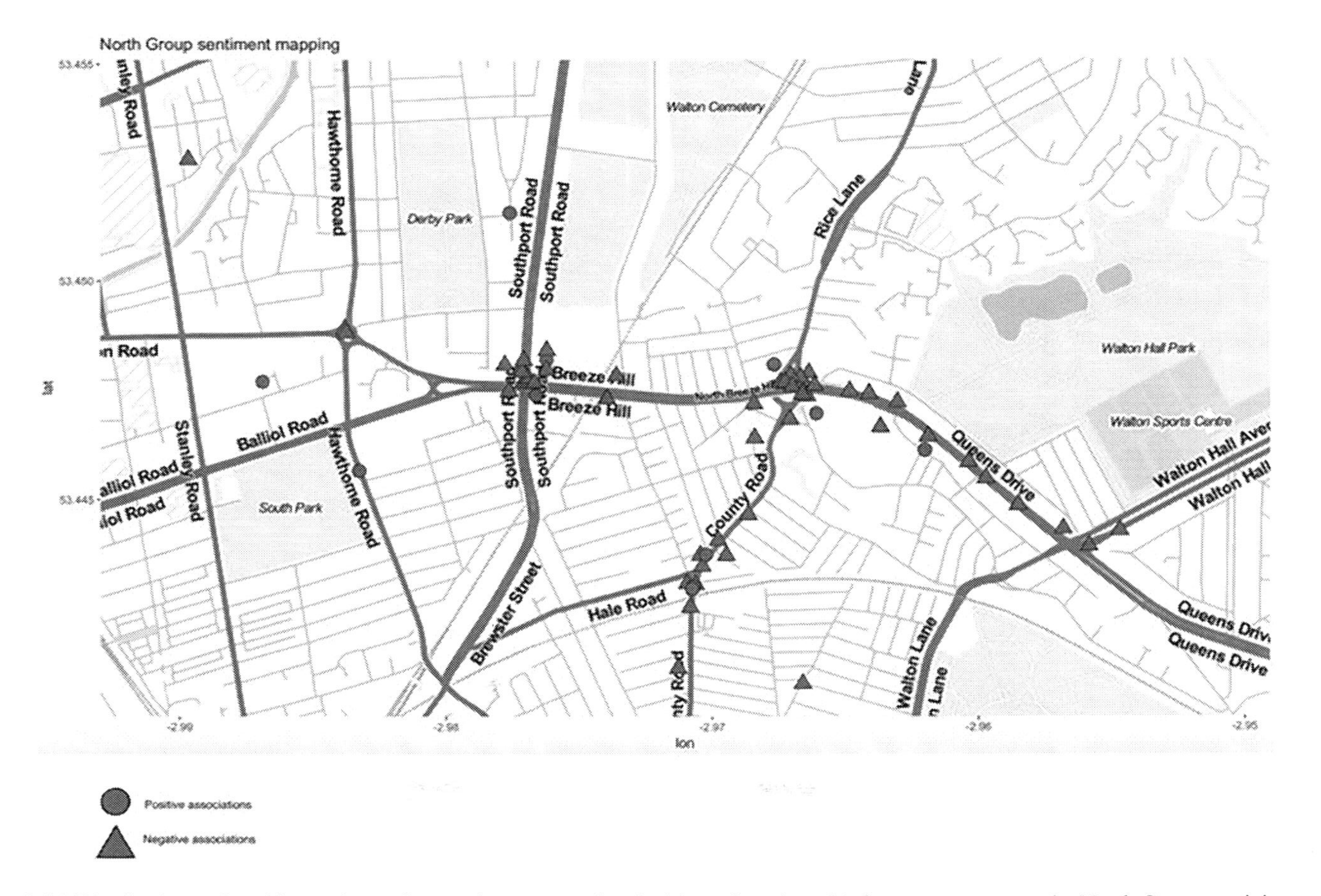

Figure 8.14 Distributions of positive and negative sentiments associated with roads and road infrastructure among the North Group participants.

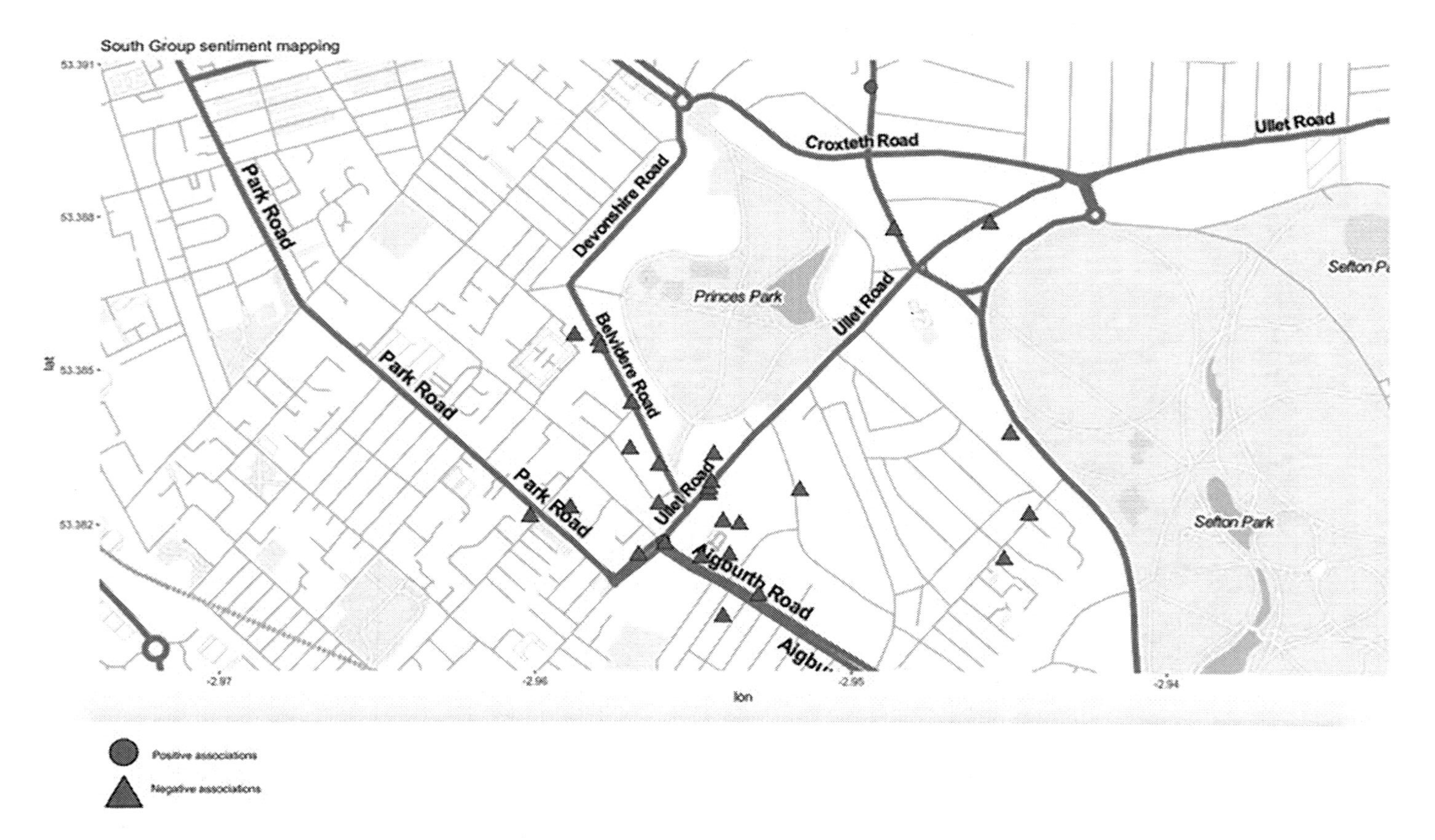

Figure 8.15 Distributions of positive and negative sentiments associated with roads and road infrastructure among the South Group participants.

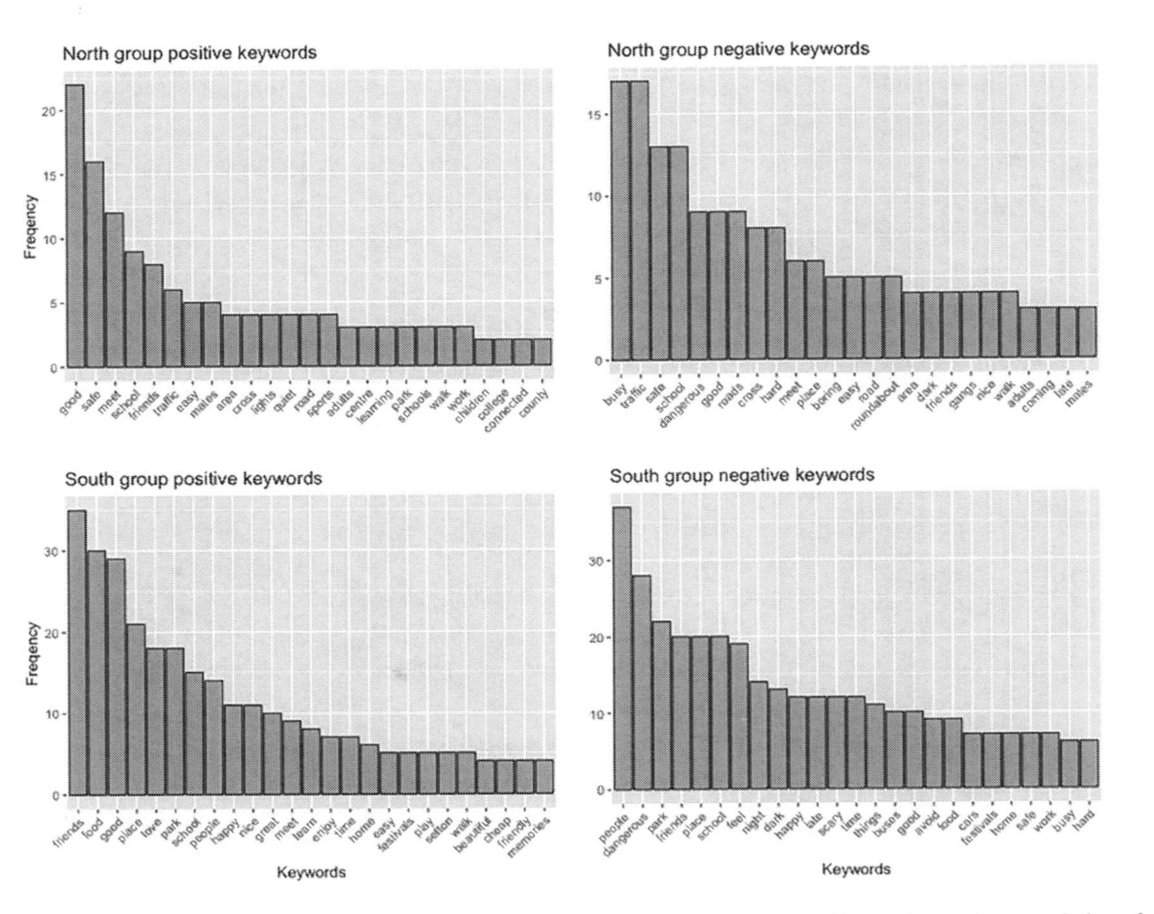

Figure 8.16 (a–d) A bar chart array representing the counts of predominant keywords with both positive and negative associations for each of the communities sampled.

'crossing'), 'lights' and so on. The dominance of a large public park in the South Group context is revealed in the close association of 'festivals', 'food', 'memories' (of a free festival) and so on. The proximity of the park also presents a range of negative sentiments associated with 'night', 'late', 'dark' and so on. The lack of negative associations with the night-time in the North Group context perhaps relates to the proximity of main roads, which provide constant light and better accessibility.

Participants in both the North and South Groups shared a common concern with being with friends in their local settings. Given this arguably natural concern for positive experiences of social life in the community context, we can also plot what assets are most closely associated with 'friends' and 'mates', including those that limit or disrupt these everyday social encounters. Figure 8.17 shows how the North Group's positive social encounters tend to happen in parkland, including around the sports centre located in a large public park. If we also refer to the plot of major features in Figure 8.8 (with points data overlaid on the network integration map), we can observe how the large park and sports centre has been selected by only one of the participant sub-groups. A major north-south road segment seems to form a boundary between these sub-groups, leaving very little opportunity for social encounter among them.

The spatial separation apparently experienced by the North Group sub-groups is experienced to a lesser extent by the South Group sub-groups. Figure 8.18 demonstrates a broader range of locations for positive social encounters among these groups. Here keywords include 'friends', 'mates', 'cafés', 'festivals', 'love', 'home', 'friendly', 'family', 'safe', 'learn', 'happy', 'socialise', 'summer', 'visit' and 'talk'. These include areas of the public parks, and also local shops and cafés. These groups also identified local bus stops as locations for positive social encounters.

Community network graph

We have now explored the spatial contexts for the sampled communities across a range of dimensions. We have been able to observe the locations of the communities in their broader urban settings. We have described the socio-economic and geographic profiles of their local spatial settings. We have been able to see how land use affects the 'shape' of the community, which is represented in a basic form by the distribution of data points. We have looked at the local road network in terms of integration centralities, showing their patterns of inter-connectivity with the wider road network. We have also seen how the proximities of these centralities influence the communities', and their sub-groups', sense of place. Certain high-value integration centralities (that is, locating busy main roads) serve to separate the community sub-groups, such that their opportunity for social encounter becomes limited.

We have also looked at the spatial contexts from the communities' perspectives, showing how both positive and negative associations are made with a range of spatial assets. We noted how the participants in each of the communities sampled

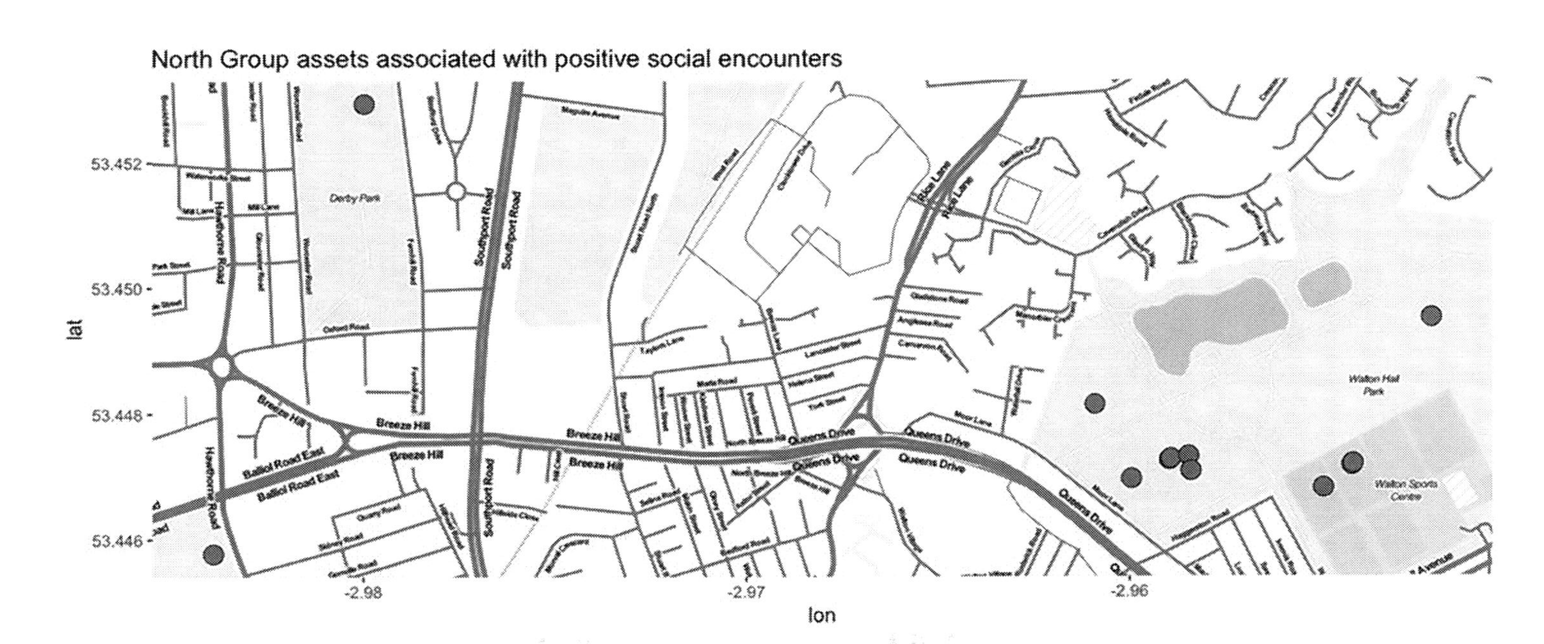

Figure 8.17 Assets associated with positive social encounters, for the North Group relating to keywords with positive associations.

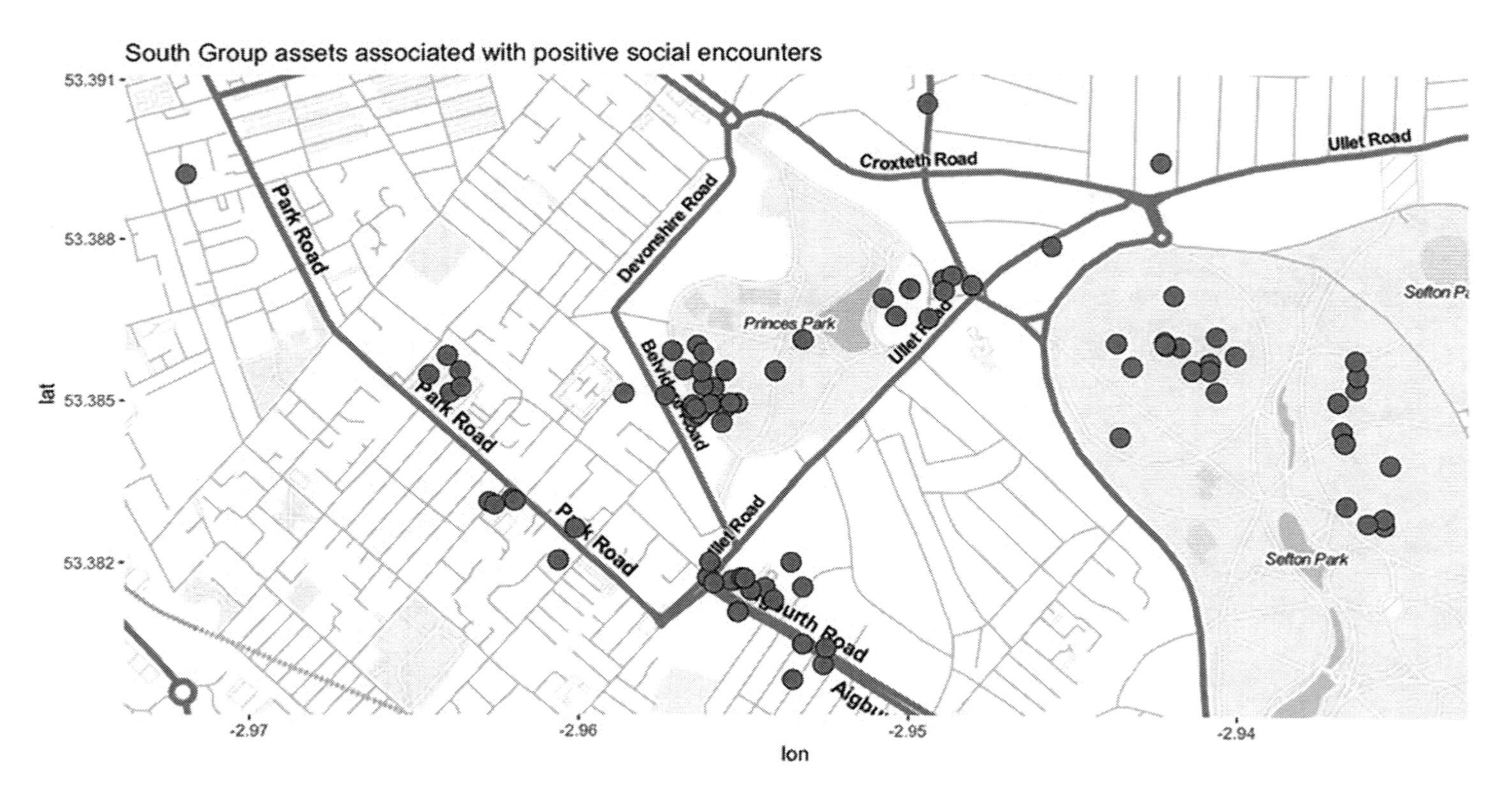

Figure 8.18 Assets associated with positive social encounters for the South Group.

highlighted roads, junctions and open space (especially public parks) as being the major features that affect their community interactions. These types of features became the focus of our investigations. We also noted how public parks feature references to positive social encounters relating to 'friends', 'family', feelings of happiness and safety and so on, as well as negative associations with 'gangs', 'people' that are not familiar and feelings of being unsafe, especially in the evening. Building on this survey of the community contexts, we can develop a network graph of relationships among the selected spatial assets. This will follow the methods for development described in Chapters 5 and 6.

The community network graph shown in Figure 8.19 represents the distribution of assets selected by the North Group, overlaid in the local section of the road network. The graph vertices are based on the centroids of points clusters, as we described in Chapter 5, and also on a set of intersections between high-integration road-network segments, which are likely to be the locations of busy junctions. These have been bound to a graph data structure, also based on the methods outlined in Chapter 5. The graph represents patterns of connectivity among the selected community assets based on their means of access. For example, the school with code 'hhs' is geographically adjacent to a sports field with code 'srpf', however, the facility can be formally accessed only via a sequence of roads and junctions. Hence, the graph represents these as a set of network steps.

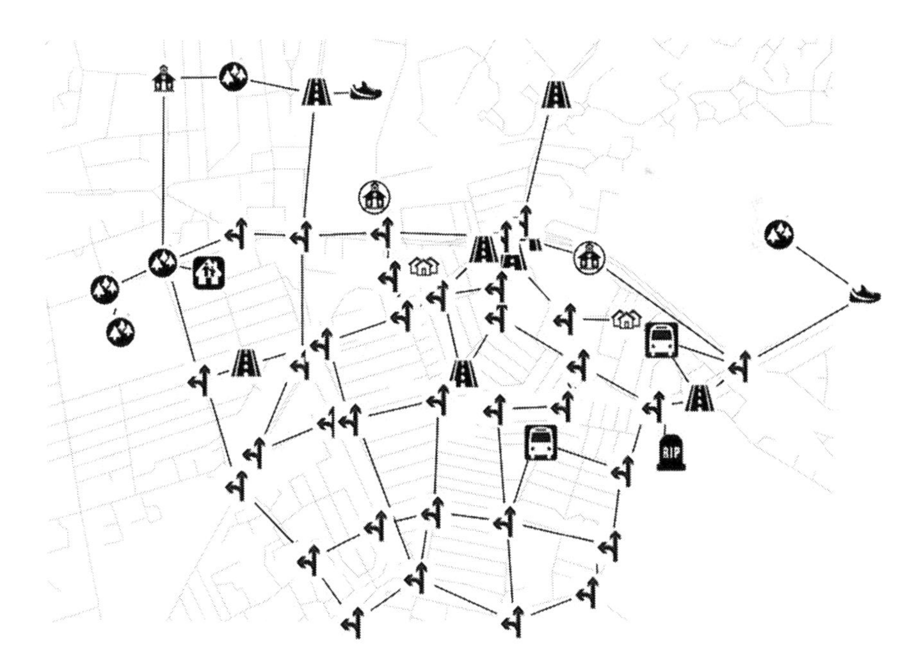

Figure 8.19 North Group community network graph with geo-located layout, plotted over street-network section.

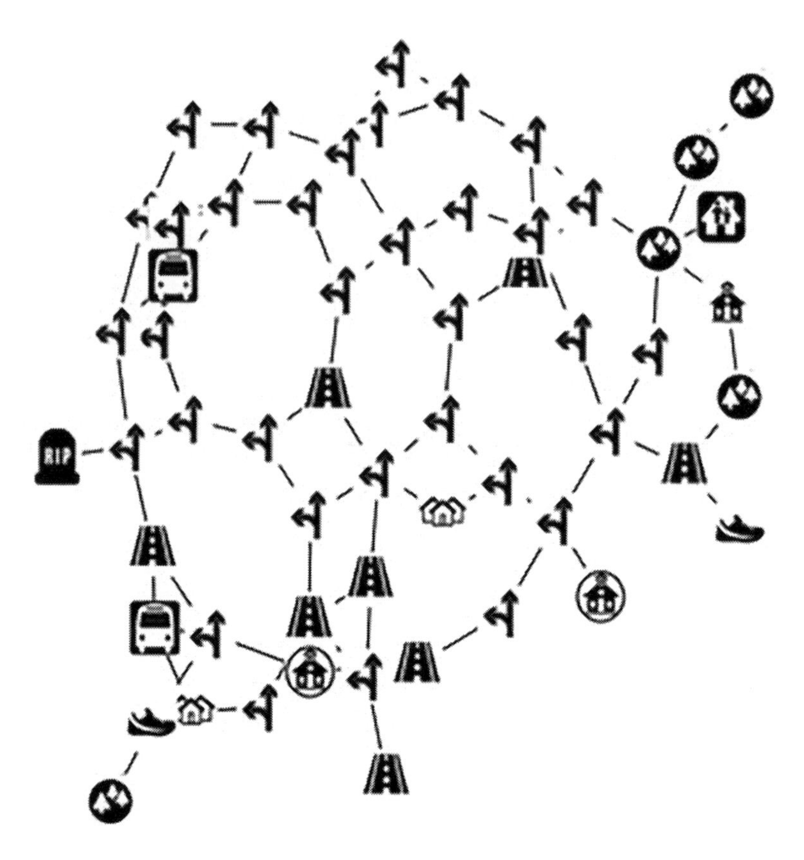

Figure 8.20 North Group community network graph with force-directed layout (Kamada Kawai format).

The community graph for North Group shows how the 'community bases' (that is, the schools where the workshops were undertaken) are connected to sub-networks of assets. We noted how the two community sub-groups that comprise North Group mainly selected assets that do not overlap. Although these groups are geographically proximate, they share very few assets in common. The community graph shows a kind of 'gap' among the vertices, which is perhaps clearer to see in Figure 8.21 as the gap between vertex 'cnr' and vertex 'j25'.

We can also plot the network based on a well-known graph layout, Kamada Kawai, which we outlined in Chapter 5. This layout in effect shows how the best-connected vertices exert a force of attraction on the graph, pushing less-connected vertices to the periphery. Looking only at the asset types, for example 'road', 'junction' or 'school', we can see how the roads and junctions constitute the set of vertices with highest centrality. Assets with type 'school', 'leisure', 'open space'

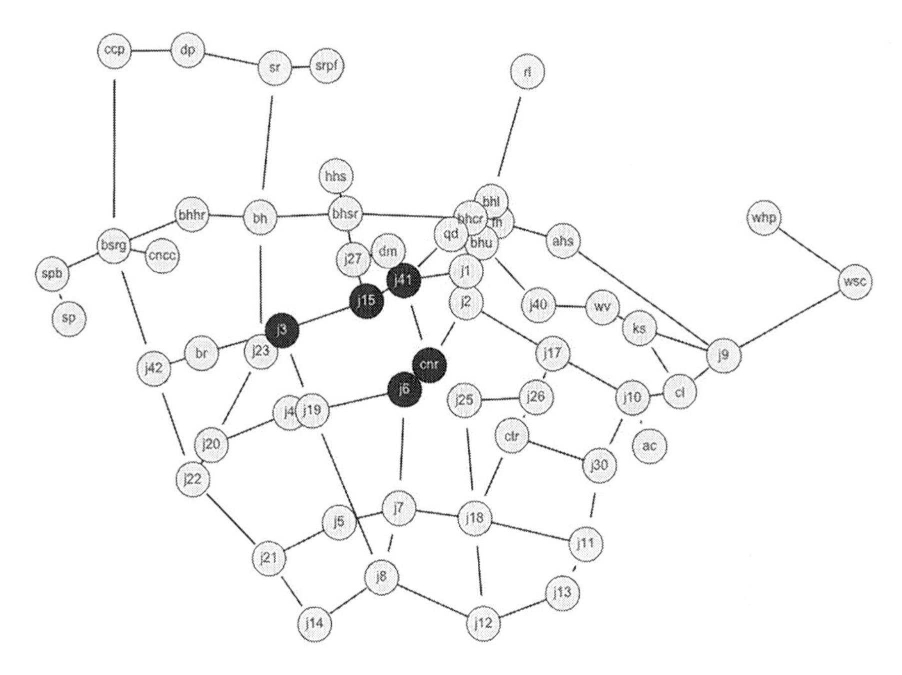

Figure 8.21 North Group community network graph with vertex labels. Vertices with highest closeness centrality values are highlighted black.

(that is, public park), 'transport', 'residential area' and 'cemetery' are pushed to the network periphery. This is possibly one spatial indicator of the separateness between the sub-groups. There appears to be no asset that is not a road or junction that is central to the community as a whole.

We can make this pattern of separation more clearly visible by calculating the closeness centralities for the vertices. Closeness represents the sum of the shortest paths from one vertex to all other vertices in the network. The higher the closeness values, the higher the sum of shortest paths connected to that vertex, and the closer it is to all other vertices. Figure 8.21 highlights in black the vertices with the top-five highest closeness centralities. Here we can see clearly how the assets with codes 'cnr' and 'j6' are central to the community network. The land-use map shown in Figure 8.21 above also reveals how a railway track runs between these assets, and it is now clear how this physical severance affects the community network at the global level. In Chapter 4 we also noted how Liverpool's inter-city railway network (the first in the world, connecting Liverpool to Manchester), was laid prior to much of the city's street layout. This seems to have had an impact on the spatial dimensions of community cohesion in some areas of the city, including in North Group's community spaces. While certain geographic or social factors may also be at play in the separation of the North Group sub-groups, we can see

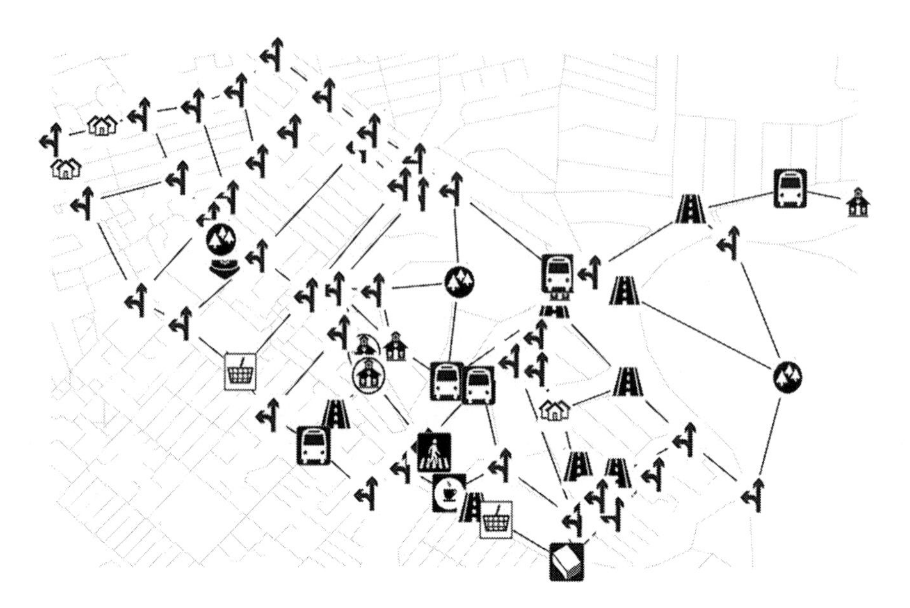

Figure 8.22 South Group community network graph with geo-located layout, plotted over street-network section.

how a spatial intervention designed to 'bridge' the sub-group community spaces could focus on the potential of building links between the assets located on vertices 'cnr' and 'j25'.

The South Group community network shown in Figure 8.22 follows a similar format to that of the North Group. The two community bases (the schools where workshops were facilitated) are geographically adjacent. As with South Group, they are also adjacent to an open space (a public park), which is accessible via two network steps. The South Group network features a variety of selected community assets, including transport hubs, a café, shops, a police station and a library. By reconfiguring the community network with a force-directed layout (Figure 8.23) we can see how the transport hubs are central to the South Group community network, along with the community bases themselves, and a public park. The other community assets of the police station, café, shops and library are pushed to the periphery of the network.

The plot of highlighted centralities in Figure 8.24 makes it clear to see how the vertices with codes beginning 'bs' (bus stops) are closest by distance to all other vertices. This is perhaps indicative of the greater degree of overlap among the community sub-groups, which we have already observed. The sub-groups of South Group have more community spaces in common, compared to the North Group sub-groups. We can also see how a subset of vertices that represent the locations of road junctions are not used in community formations.

Figure 8.23 South Group community network graph with force-directed layout (Kamada Kawai format).

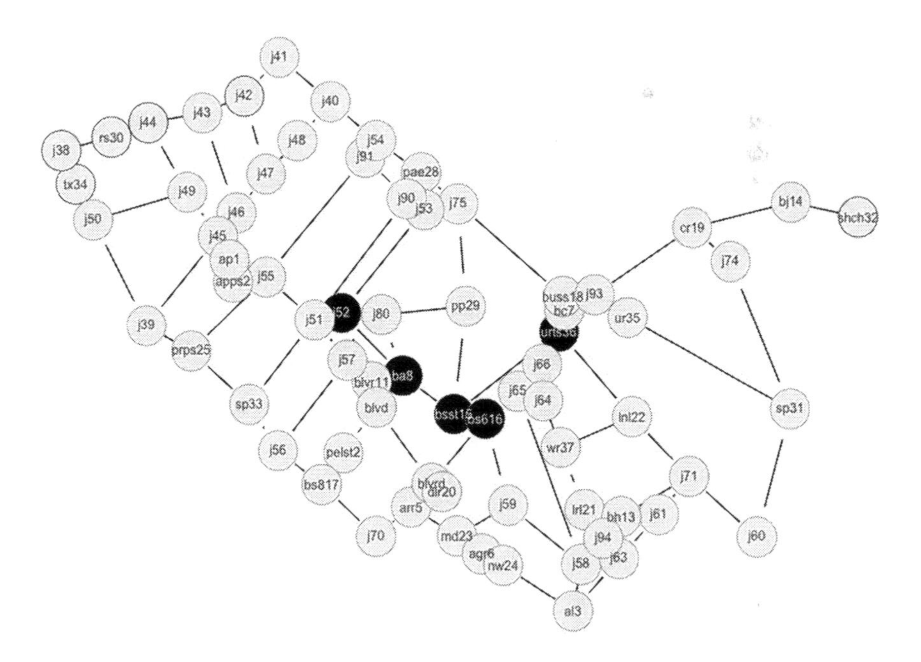

Figure 8.24 South Group community network graph with vertex labels. Vertices with highest closeness centrality values are highlighted black.

In fact, they are pushed to the periphery of the network. Overall the sub-groups that make up this community formation appear to experience a range of assets at the centre of their community network. A design intervention in the community space could perhaps focus on bringing some assets at the community periphery, such as the library at vertex with code 'al3', into the network centre. This could be achieved by establishing 'linking' or 'bridging' mechanisms, which might, for example, comprise a lit pathway or improved crossings at vertex (junction) 'j59'.

The basic closeness graphs presented in this section serve to show how certain vertices might converge paths across the network. As such, they represent key locations for community formations, or at least places where people encounter each other. It can also be useful to have key community assets based not at the very centre of the network, but close enough to it. This means that the asset might be protected from the wider-scale movements across the network, involving people not directly connected to the area. To some degree, they form the social and spatial contexts for community life, even if they were not designed with reference to locally embedded parameters (as in the case of globally branded shops in local settings). In other words, the graph position of these assets helps to locate what could be termed 'defensible spaces' for community formations (Newman, 1996).

Directed graphs

Centralities are functions of network structure, measured by distances among its vertices and edges. However, centrality does not show how the community located across this network makes use of these centralities. Centralities can both attract and repel based on their network functionalities, and also on certain social or cultural factors at play in the community spaces. We noted in Chapter 2 how informal settlements in Brazil ('favelas') present centralities that attract people in the day for social encounter and commerce, yet repel them from the afternoon onwards due to illegal gang activities.

Community perspectives on network formations can be represented as a directed graph, in which vertices bear network weightings on unequal terms. Some have more weight, some less, or some have positive influence and some negative influence. Following the methods for generating weighted and directed graphs outlined in Chapter 6, we can make use of the participants' text descriptions of their community spaces to evaluate vertices with overall positive or overall negative associations. This provides us with a 'sketch' of attraction from the community's point of view, which should not be read as a valid or naturalistic model of their movements, only an indicator for grounded investigation. As such, the weighting mechanism is based on a kind of voting system stemming from the evidence of the participatory materials, in which vertices 'benefit' from the participants' descriptions, or else from their connectivities to those vertices.

The directed graph for North Group (Figure 8.25) presents a mixed pattern of attraction. This involves non-converging attraction that does not seem to settle

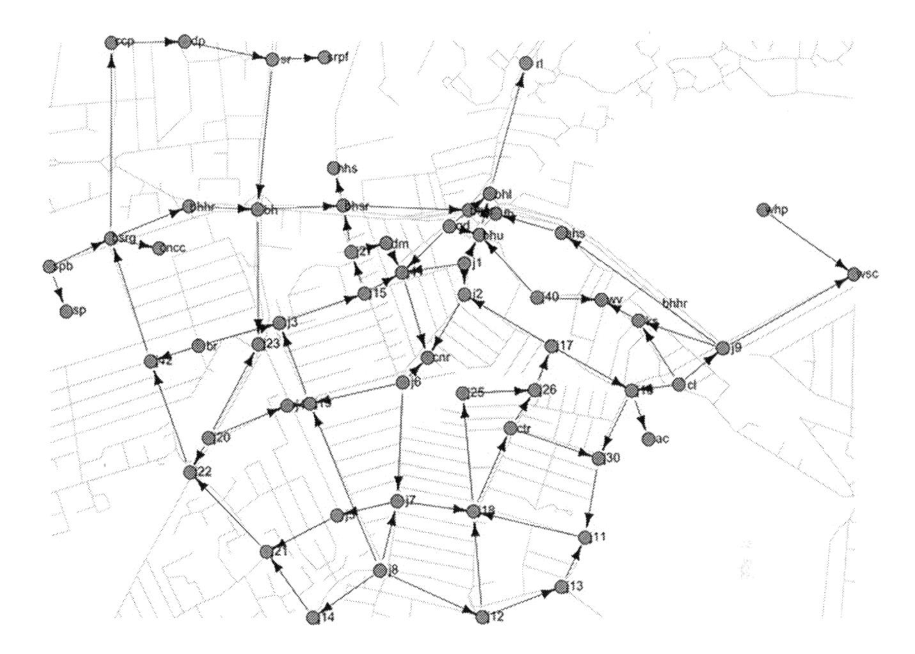

Figure 8.25 Directed graph of the North Group community network.

on specific vertices, but is pushed from one vertex to another. It also involves some isolated examples of attraction converging on vertices. For example, attraction seems to be reinforced on the high closeness centrality of 'cnr': its network functionality seems to relate to community attraction. While vertex 'j25' is close to 'cnr' (in terms of Euclidean, 'as the crow flies' distance), it is not connected to it within the network, and has not benefitted from its high level of attraction. Other attractors from the community perspective occur around a major road intersection, for example the vertex 'bhu', perhaps relating to the crossing places used by many of the participants to each of their community bases (their school buildings). Some other vertices, such as 'j41', are not central to the network graph, however they are useful assets from the community perspective, perhaps because they form intersections of pathways that bypass the high-integration segments (the busy main roads).

The directed graph for South Group (Figure 8.26) also reinforces some network centralities from the community perspective. We can see these, for example, on the vertex with code 'urts36', which is the location of a transport hub, and 'sp31', a public park. As with the North Group directed graph, we find here that the patterns of attraction converge on only a small number of vertices. Some, including the vertex 'al3', the location of a local library, should perhaps converge greater attraction than is currently the case. This particular asset lies at the functional

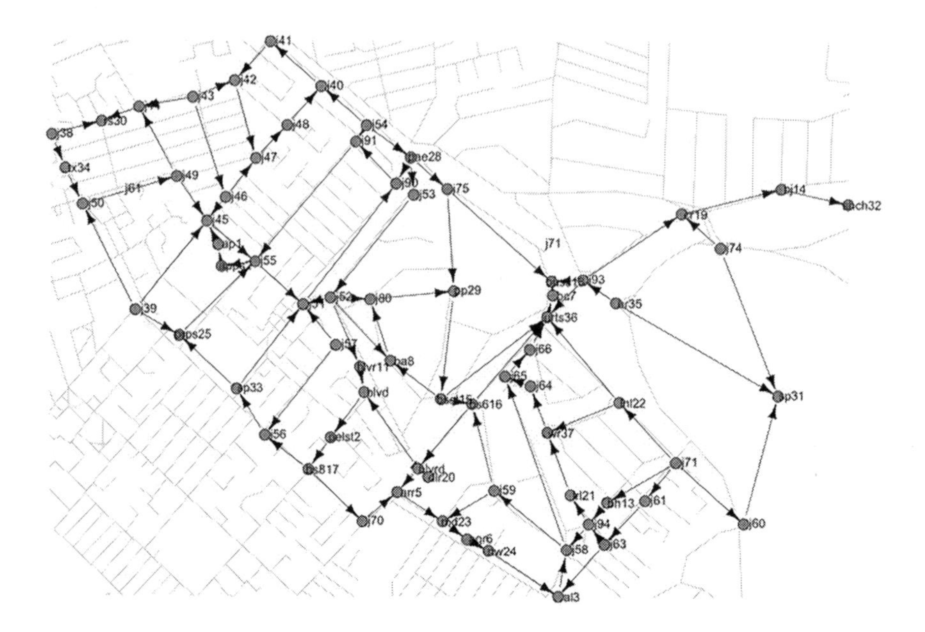

Figure 8.26 Force-directed graph of the South Group community network.

edge of the community network, and its attractiveness could perhaps be increased through design interventions.

Conclusion

Making use of the series of maps and network models presented in this chapter has allowed us to observe the contexts for community formations in each of the case-study settings. We have seen how each community is located at the edge of Liverpool's urban centre, within zones presenting moderate to high deprivations. These multiple deprivations appear to be mitigated somewhat by overall good accessibility to key community assets, such as doctor's surgeries or general stores. We also looked at the impact of non-residential land use on community formations and were able to observe how the sampled communities form largely within residential zones (including locations of local shops). We observed how some entirely inaccessible areas, comprising, for example, the routes of railway lines or sites of industrial zones, serve as hard boundaries for community formations. Railways can also bisect community spaces, as we observed in the case of the North Group.

Observations of physical limitation do not always include the range of other factors that might affect accessibility, including experiences of safety and security and perceptions of boundaries or other conceptualized limitations. For this reason, we drew from community perspectives of the case-study settings to reveal different kinds of attraction within the community network. We observed how the communities sampled focus their 'sense of place' around certain key assets, including roads and junctions, open spaces such as public parks, and some others including cafés, libraries, or sports and leisure centres. We focused our attention on the major asset types of roads and junctions, and open spaces. We noted how these were selected by the participants based on both negative and positive connotations. We extracted sets of keywords associated with these connotations and, perhaps unsurprisingly (maybe even reassuringly), found that the greater factors in positive experiences were attributable to friendships and the ease by which these could be enjoyed. Factors relating to traffic and unfamiliar people were attributable to bad experiences. Public parks also received highly mixed associations in this regard, with positive associations relating to organized events in these spaces. Negative associations with public parks seemed to relate to the presence of unfamiliar people and to gangs, along with risks or anxieties associated with evening or night-time darkness.

We sought to determine what assets lie at the heart of each community, in terms of both network functionalities and community attractors. To achieve this, we generated models of network centrality based on measures of closeness, calculated as the sum of the shortest paths connecting a vertex to each other vertex. This revealed contrasting contexts for the community formations, as the North Group's key assets were all pushed to the network periphery, whereas many of the South Group's key assets were located at the functional centre of the network. We also noted that having assets within the network centre can mean that they become accessible to the network at the global level, meaning they may be less 'defensible' relational to locally embedded parameters. In other words, they are open potentially to all comers, and are not reserved for one particular group. The merits of having assets at the centre or periphery of the network are dependent on the local community contexts. It appears that a café used by the South Group is located at the periphery of the community network, and remains a beneficial place for community life. Some areas of a local park are similarly peripheral to the community network, and appear to be associated with risks or anxieties from unfamiliar people and gangs. The centrality of local transport hubs seems to be highly beneficial to each of the South Group sub-groups (the two schools where the workshops were facilitated).

Overall, we might be struck by the paucity of spatial assets for the communities sampled. There appear to be few places for participants to meet members of their own community, or encounter other people in a positive light. The community spaces are bounded or bisected by infrastructural assets, such as roads and railways. Accessibility to potentially beneficial assets, such as nearby parks and leisure facilities, is hampered by roads, lack of direct entrance points, poor lighting and so on. There appears to be very little support for organic connectivity between the assets selected by the community.

We can highlight examples in peripheralized assets in the North Group and the South Group. The North Group features a sports facility, which is used after school by many of the participants. It is just about the only asset in the community space at which members of both sub-groups might encounter each other in a proactive capacity (other assets include more socially 'passive' spaces such as roads and junctions). Yet this key asset is pushed to the periphery of the community network by poor accessibility. It is reached through a single entrance to a park, rather than directly from various points along the road network. This of course might also mean it is defensible – not open to all comers – but its benefit to community formation might be overlooked. For the South Group community, the example of the public library being at the periphery of their community network is perhaps testament to the oversights in path-finding in this context. A public library is a space where young people can meet after school, or do some work, in a quiet, safe, comfortable, supervised environment. It is a space that might help them to negotiate the transition from the dependencies of home and school life, to becoming independent in urban settings. We might examine in greater depth the reasons why these essential assets, being rare benefits to community formation in the spaces we sampled, remain at the peripheries of their participants' networks.

Epilogue

Design research seeks to observe, analyze and understand the contexts in which habitable forms take shape. At a fundamental level, there is no form without context. Our perceptions of the world around us, and our place in it, are based on the processing of information drawn from the environment. We distinguish the properties of a form against those of its environment; we seek out the differences among things, and describe these in space: 'differences are the things that get into a map' (Bateson, 2000, p. 459). Contexts might be physical, cultural, political or ideological. The contexts in which urban communities form comprise the physical structures of streets, buildings and open spaces. Their cultural contexts comprise locally embedded conventions, or subtle systems of reward and penalty, some of which are common to everyone, and some of which are bound to a specific place. Contexts involve decision-making processes, often hierarchical in nature, sometimes lead to factional conflicts and the seeking of compromises and resolutions. They might also involve the enforcement of formalized rules and mores, envisioned through iconic, heroic or monumental imagery.

Members of urban communities, whether individuals or sub-groups, negotiate their personal and social identities via their contextual conditions and parameters. They use structures, such as their dwelling or a nearby street corner, to make, to share and to memorialize their sense of place: to tell 'stories of us' or, conversely, 'stories of them'. There are stories of places where they were helped and protected, where they encountered fear and danger, where they met friends, where they encountered strangers. Stories always belong to places, and places unfold through storytelling. Places are memories, and stories help us to share these memories over time.

To understand the contexts in which habitable forms take shape, the field of design research must engage in the stories of places. Storytelling is not always verbal, it is often performed through everyday activities. Sometimes we linger in one place, we stand still quietly and look about, we gaze through a café window, for no declared purpose but to be a gazer. We look askance or speak angrily to an obstacle in the street that will not respond to us, at least not verbally. We expect our environment to engage in a kind of thinking in the sense that it yokes things together into a continuous chain of meanings. Our street links to a street corner, which concatenates to another street, to a main road, linking to squares and open spaces. This self-similarity is reflective of a natural order, as universal as gravity or entropy, which stems from interactions of simple elements and builds towards complex forms that are not describable in terms of their simple parts alone. Our thinking follows this pattern, deriving 'atomic' information from the environment, sketching out its basic forms, then attending to its textures and dynamics. Yet we cannot explain our thinking through a disaggregation of its components. Our thinking always adds something to the world: a story. We 'think with' the self-similar patterns of our environment, we tell our stories through its pathways and dead-ends, its sunlight and shadows. Our worlds are storied landscapes (Ingold, 1993).

The chapters of this book sought to develop a toolkit for capturing these stories of the built environment, both as verbal or textual descriptions, and as descriptions in space. I developed my thinking about design research based on the foundational work of, respectively, Lefebvre (1991), Simon (1996), Alexander (1971) and Hillier and Hanson (1984). These authors have each demonstrated ways in which the complexity in the built environment unfolds from interactions among basic elements, and that urban forms comprise hierarchical systems. I drew from the seminal work of Lynch (1960), who described ways in which we navigate our environment by means of its components. I also drew from Hillier (2007) the notion of 'relational complexity' to describe how our interactions with components in the built environment bring about conceptual-physical combinations in the artefacts of our everyday worlds. In summary, our worlds are unfolding landscapes, never settled and continuously transformed through our thinking: our storytelling about what was, what is, and what should be.

I have focused on the development of a toolkit based on the case studies of community formation in the city of Liverpool. Liverpool was one of the first cities in the 19th century to grow rapidly from a small port into a major maritime city. The urban patterns that stem from this rapid growth are today very familiar, so much so that we may not recognize them as having special properties that are different to those of a rural village. The mosaic patterns of rectilinear, convex and concave forms bear significance for their particular inhabitants' sense of place: their centralities, connectivities and boundaries, their community locales, ways to go and sense of safety or danger. Observing the spatial evolution of the city's road network helped to place our communities' formations in time as well as space.

I outlined a set of computational tools, written in the R programming language, that might help us to describe community formations. Crucially, I aimed to bind descriptions to their locations, and I made use of a graph data structure to achieve this. This structure also allowed us to decouple the locational descriptions from their geo-coordinates, and to observe them through the patterns of their network distances. Hopefully this has served to demonstrate both the spatial and the semantic dimensions of community formations. I have also endeavoured to show how communities exert different forces over the network, pulled towards certain kinds of attractors, or pushed away from others. This was an experimental exercise and some further development in this area might help to describe these directional flows to enrich our decision-making for designing and planning.

Developing an experimental theme, I also adopted an iconographic lexicon for describing spatial relationships. In reviewing the case-study materials, it may be possible to describe the community formations observed by way of a self-similar and portable set of spatial descriptions. My intention was not to reduce the participants' descriptions to a basic spatial schema, but to configure a tool for design thinking based on simple-to-complex relationships.

This book has taken design-research a methodological approach to spatial complexity in urban community formations. I have drawn from a 'configurational' approach to spatial morphology, as reflected in the theory and method of space syntax. This theory and method regards urban spatial structures as being reflective of the social patterns and processes that take shape around them. Societies configure spaces to reflect their persistent structures and norms, including the need for accessibility and the need for privacy. This approach also allows us to observe the evolution of spatial structures over time, especially as the syntactical network (the self-similar, hierarchical patterning of minor and main roads and their intersections) is 'normalized'; that is, it does not include historical or geographic information, meaning that network structures can be compared like-for-like from one period of time to the next (Vaughan, 2018, p. 17).

The configurational approach of space syntax also allows design researchers to consider the *affordances* of network structures and their potential to fulfil their design function. Affordances may be realized through deliberate actions following declared sets of rules or protocols, or through accidental disruptions. In the latter case, think of the 'desire paths' that form over grass lawns and verges, worn out through people's to-ing and fro-ing. The potential for these desire lines is a *latent* property of the global network structure. This property emerges from people's interactions (expressed in movements) along the formal structures of the urban road network, which directs flows of movement along predictable patterns. Where these patterns come up against the edges of open spaces, so the movements they directed have the potential to continue, insofar as they have a reason to and are able to keep moving. People will follow the path of least effort where this is available to them. They will cut across the grass rather than go around the edge. The patterns of the formal network often come to follow these informal, pedestrian path-finding processes.

Potential affordance in the shaping of townscapes is, arguably, an under-utilized notion in design research. Our design thinking is very often path-dependent, following sets of prior knowledge and established protocols, as of course it has to be in order to deliver portable solutions to design problems. Yet our towns are facing unprecedented challenges to their resilience and sustainability. Many grew from agrarian lifeways, from seasonal cycles, the need to barter and the need to move by foot or cart. Many were transformed radically in the 19th century around expansions in mining and manufacturing, reflected in densified terraced housing (which successfully persists to this day). Many experienced a further, radical disruption in response to the widespread adoption of mass transit, leading to the proliferation of low-density suburban housing (also successfully persistent to this day). Later, the mainstream adoption of the private automobile, coupled to centralism of post-war planning, led to the building of city-scale infrastructures – orbitals, arterials and interchanges – to form voids and hard edges in the urban fabric, often serving as lacunae and boundaries in community life.

Waves of expansion and segmentation have transformed our townscapes, yet they have remained spatially resilient to social, cultural and technological changes. The challenges they face are economic and political. Economic challenges stem from the dismantling of major industries in the late 20th century. They are political because the dismantling was undertaken without a practicable model for redressing the decline in opportunities for employment. The towns that expanded rapidly around mining, manufacturing or commerce were stripped of their economic lifelines. While many industrial working-class families experienced poverty and hardship when industrial capacity was at its maximum, they came to be denied the possibility for generational improvement to their conditions (Froud et al., 2012). Many towns have become less viable for well-supported family life, and this is reflected in the years-long decline in their working-age populations. Many towns suffer low business density and retail occupancy; high streets feature rows of closed and boarded-up shop fronts. Public services such as social care and transport have become privatized and residual.

Yet the spatial resilience of towns, and of the cities that came to encompass them, means that their potential to be assets for community life remains intact. Former market towns remain strategically located in relation to other towns, cities and village settlements. The needs of vitality and welfare at the regional level that sustained their early growth remain prominent; people still need to enjoy a good life, to be connected, to work happily, to be cared for. There is increasing evidence for the negative effects of public management models that have sought to operationalize community needs, arguably with a focus on business performance rather than on personal and public welfare.[1] A response in public provision to these negative effects has been made through the local procurement of services from within a network of 'anchor' partnerships (Bentham et al, 2013).[2] This refocusing on local delivery and accountability is reflective of what might be called the organic inter-dependencies of community life: organic because they are cohesive and possibly self-regulating systems, and inter-dependent because they have some needs in common, as do all people and places, yet they work best when local knowledge is brought into thinking around local problems. There is no organic community that has everything it needs all together in one place.

Each urban community, whether in city, town or village, has the potential to offer something unique: something they do particularly well, whether due to historical accident, environmental advantage, well-honed craft or otherwise. Or else they can offer a unique opportunity for others to help; they might lack a resource, or can articulate a set of requirements, that others can meet for their improvement. Towns, and townscapes within cities, even when beleaguered by decline, can nevertheless become active components of their regional network, to capitalize and to share the benefits of their inter-connectivities, if only these inter-connectivities can be reimagined as available assets.

The challenge for a new wave of design thinking around urban community formation stems from the difficulties of imaging their real or potential inter-dependencies within their spatial settings. The methods for connective modelling outlined in this book help in the graphic description of community spaces as 'relational artefacts', involving their inhabitants' uses of physical and conceptual objects to form their community identities. These methods might offer the possibility of a toolkit for design research that serves to reimagine community spaces not as drains on finite resources but as assets for welfare.

Notes

1 For example, "How outsourcing fell out of fashion in the UK", *Financial Times*, 9th February, 2018. Author: Gill Plimmer. Available via ft.com (accessed 20 March 2019).
2 For example, the 'Preston Model': www.preston.gov.uk/thecouncil/the-preston-model/preston-model/ (accessed 20 March 2019).

References

Alexander, C. (1971). *Notes on the synthesis of form*. Paperback ed. Cambridge, MA: Harvard University Press.

Alexander, C., Ishikawa, S., Silverstein, M., Jacobson, M., Fiksdahl-King, I. and Angel, S. (1977). *A pattern language: Towns, buildings, construction (Center for Environmental Structure Series)*. New York, NY: Oxford University Press.

Al-Sayed, K. (2014). Thinking systems in urban design: A prioritised structure model. In: M. Carmona, ed., *Explorations in urban design*. Farnham: Ashgate, pp. 169–181.

Al-Sayed, K., Turner, A. and Hanna, S. (2009). Cities as emergent models: The morphological logic of Manhattan and Barcelona. In: D. Koch, L. Marcus and J. Steen, ed., *Proceedings of the seventh international space syntax symposium*. Stockholm: KTH Royal Institute of Technology.

Amin, A. and Thrift, N. (2002). *Cities: Reimaging the urban*. Cambridge: Polity Press, pp. 78–79.

Anciaes, P., Boniface, S., Dhanani, A., Mindell, J. and Groce, N. (2016). Urban transport and community severance: Linking research and policy to link people and places. *Journal and Transport & Health*, 3, pp. 268–277.

Aughton, P. (2008). *Liverpool: A people's history*. 3rd revised ed. Lancaster: Carnegie Publishing.

Bateson, G. (2000). *Steps to an ecology of mind: Collected essays in anthropology, psychiatry, evolution, and epistemology*. New ed. Chicago: University of Chicago Press.

Batty, M. (2013). *The new science of cities*. Cambridge, MA: MIT Press.

Bentham, J., Bowman, A., de la Cuesta, M., Engelen, E., Ertürk, I., Folkman, P., Froud, J., Johal, S., Law, J., Leaver, A., Moran, M. and Williams, K. (2013). *Manifesto for the foundational economy*. CRESC Working Paper No. 131. University of Manchester: Centre for Research on Socio-Cultural Change (CRESC)

Berman, T. (2017). *Public participation as a tool for integrating local knowledge into spatial planning: Planning, participation and knowledge*. Cham, Switzerland: Springer International Publishing AG.

Berta, M., Caneparo, L., Montouri, A. and Rolfo, D. (2016). Semantic urban modelling: Knowledge representation of Urban space. *Environment and Planning B: Planning and Design*, 43(4), pp. 610–639.

Brand, S. (1997). *How buildings learn: What happens after they're built*. Revised ed. London: Phoenix Illustrated.

Cassiers, T. and Kesteloot, C. (2012). Socio-spatial inequalities and social cohesion in European cities. *Urban Studies*, 49(9), pp. 1909–1924.

Chambers, R. (1995). Poverty and livelihoods: Whose reality counts? *Environment and Urbanization*, 7(1), pp. 173–204.

Chambers, R. (1997). *Whose reality counts? Putting the first last*. London: Intermediate Technology Publications.

Chein, M. and Mugnier, M. L. (2009). *Graph-based knowledge representation: Computational foundations of conceptual graphs*. London: Springer-Verlag.

Childers, D., Pickett, S., Morgan Grove, J., Ogden, L. and Whitmer, A. (2014). Advancing urban sustainability theory and action: Challenges and opportunities. *Landscape and Urban Planning*, 125, pp. 320–328.

Coleman, D., Georgiadou, Y. and Labonte, J. (2009). Volunteered geographic information: The nature and motivation of producers. *International Journal of Spatial Data Infrastructures Research*, 4, pp. 332–358.

Conroy Dalton, R. and Bafna, S. (2003). The syntactical image of the city: A Reciprocal definition of spatial elements and spatial syntaxes. In: *Proceedings of the fourth international space syntax symposium*. London.

DCLG. (2015). *English indices of deprivation 2015*. London: Department of Communities and Local Government.

Debertin, D. and Goetz, S. (2013). *Social capital formation in rural, urban and suburban communities*. University of Kentucky: Staff Paper 474, Oct. 474.

Dennis, R. (2008). Urban modernity, networks and places. In: *History in focus: The city*. Available at: www.history.ac.uk/ihr/Focus/City/articles/dennis.html

Doorst, K. and Dijkhuis, J. (1995). Comparing paradigms for describing design activity. *Design Studies*, 16, pp. 261–274.

Ellen, I. and Turner, M. (1997). Does neighborhood matter? Assessing recent evidence. *Housing Policy Debate*, 8(4), pp. 833–866.

Engels, F. (1884/1991). Socialism: Utopian and scientific. In: Karl Marx and Frederik Engels, eds., *Selected works*. Revised ed. London: Lawrence and Wishart.

Farías, I. and Bender, T. (2010). *Urban assemblage. How actor-network theory changes urban studies*. Abingdon: Routledge.

Forty, A. (2004). *Words and buildings: A vocabulary of modern architecture*. London: Thames & Hudson.

Froud, J., Johal, S., Moran, M. and Williams, K. (2012). Must the ex-industrial regions fail? *Soundings*, 52(52), pp. 133–146.

Gans, H. (2002). The sociology of space: A use-centered view. *Cities & Communities*, 1(4), pp. 329–340.

Gans, H. (2006). Jane Jacobs: Toward an understanding of "Death and Life of Great American Cities". *Cities & Communities*, 5(3), pp. 213–215.

Gibson, J. (1979). *The ecological approach to visual perception*. Hillside, NJ: Lawrence Erlbaum Associates.

Grannis, R. (1998). The importance of trivial streets: Residential streets and residential segregation. *American Journal of Sociology*, 103(6), pp. 1530–1564.

Grannis, R. (2005). T-communities: Pedestrian street networks and residential segregation in Chicago, Los Angeles, and New York. *City & Community*, 4(3), pp. 295–321.

Grannis, R. (2009). *From the ground up: Translating geography into community through neighbor networks*. Princeton, NY: Princeton University Press.

Griffiths, S. (2009). Persistence and change in the spatio-temporal description of Sheffield Parish c.1750–1905. In: D. Koch, L. Marcus and J. Steen, eds., *Proceedings of the seventh international space syntax symposium*. Stockholm, Sweden: Royal Institute of Technology (KTH).

Griffiths, S. (2011). Temporality in Hillier and Hanson's theory of spatial description: Some implications of historical research for space syntax. *Journal of Space Syntax*, 2(1), pp. 73–96.

Griffiths, S. (2012). The use of space syntax in historical research: Current practice and future possibilities. In: M. Greene, J. Reyes and A. Castro, eds., *Proceedings of the eighth international space syntax symposium*. Santiago de Chile: PUC.

Griffiths, S. (2013). GIS and research into historical "spaces of practice": Overcoming the epistemological barriers. In: A. von Lünen and C. Travis, eds., *History and GIS: Epistemologies, considerations and reflections*. London: Springer, pp. 153–172.

Griffiths, S. (2017). Manufacturing innovation as spatial culture: Sheffield's cutlery industry c.1750–1900. In: I. Van Damme, B. Blondé and A. Miles, eds., *Cities and creativity from the renaissance to the present*. Abingdon: Routledge.

Gwyther, G. (2005). Paradise planned: Community formation and the master planned estate. *Urban Policy and Research*, 23(1), pp. 57–72.

Haklay, M., Antoniou, V., Basiouka, S., Soden, R. and Mooney, P. (2014). *Crowdsourced geographic information use in government*. Report to the Global Facility for Disaster Reduction and Recovery (GFDRR). London: World Bank.

Hanley, L. (2017). *Estates: An intimate history*. New ed. London: Granta Books.

Hanson, J. and Hillier, B. (1987). The architecture of community: Some new proposals on the social consequences of architectural and planning decisions. *Architecture et Comportement/Architecture and Behaviour*, 3(3), pp. 251–273.

Hay, R. (2011). *Guide to localism opportunities for architects: Part two: Getting community engagement right*. London: Royal Institute of British Architects.

Heinrichs, D. (2016). Autonomous driving and urban land use. In: M. Maurer, J. Gerdes, B. Lenz and H. Winner, eds., *Autonomous driving*. Berlin, Heidelberg: Springer

Hillier, B. (1999). Centrality as a process: Accounting for attraction inequalities in deformed grids. In: *Proceedings of the second international space syntax conference*. Brasilia, Brazil.

Hillier, B. (2007). *Space is the machine*. Electronic ed. Cambridge: Cambridge University Press.

Hillier, B. and Hanson, J. (1984). *The social logic of space*. Cambridge: Cambridge University Press.

Hillier, B. and Iida, S. (2005). Network and psychological effects in urban movement. In: A. Cohn and D. Mark, eds., *Spatial information theory*. COSIT 2005. Lecture Notes in Computer Science, Vol. 3693. Berlin: Springer-Verlag.

Hillier, B., Turner, A., Yang, T. and Park, H. T. (2010). Metric and topo-geometric properties of urban street networks: Some convergences, divergences and new results. *The Journal of space syntax*, 1(2), pp. 258–279.

Hillier, B. and Vaughan, L. (2007). The city as one thing. *Progress in Planning*, 67(3), pp. 205–230.

HMSO. (1981). *The Scarman report: Report of an inquiry by the Rt. Hon. The Lord Scarman, OBE*. London: Her Majesty's Stationery Office.

Hoadly, C. and Cox, C. (2009). What is design knowledge and how do we teach it? In: D. DiGiano, S. Goldman and M. Chorost, eds., *Educating learning technology designers: Guiding and inspiring creators of innovative educational tools*. Abingdon: Routledge, pp. 19–35.

Hobsbawm, E. (1962). *The age of revolution: Europe 1789–1848*. London: Weidenfeld and Nicholson.

Hommels, A. (2010). Changing obdurate urban objects. In: I. Farías and T. Bender, eds., *Urban assemblages: How actor-network theory changes urban studies*. Abingdon: Routledge, Chapter 6.

Hoskins, G. W. (1955/2013). *The making of the English Landscape (Nature Classics Library)*. Dorchester: Little Toller Books.

Ingold, T. (1993). The temporality of the landscape. *World Archaeology*, 25(2), *Conceptions of Time in Ancient Society*, pp. 152–174.

Ingold, T. (2000). *The perception of the environment: Essays on livelihood, dwelling and skill. Chapter 12: Globes and Spheres: The topology of environmentalism*. London: Routledge.

Ingold, T. (2011a). Against space: Place, movement and knowledge. In: T. Ingold, ed., *Being alive: Essays on movement, knowledge and description*. Abingdon: Routledge.

Ingold, T. (2011b). *Being alive: Essays on movement, knowledge and description. 'Epilogue: Anthropology is NOT ethnography'*. Abingdon: Routledge.

Innes, J. and Booher, D. (2004). Reframing public participation: Strategies for the 21st century. *Planning Theory and Practice*, 5(4), pp. 419–436.

Jansen, H. (1996). Wrestling with the angel: On problems of definition in Urban historiography. *Urban History*, 23(3), pp. 277–299.

Johnson, J. (2010). Embracing design in complexity. In: K. Alexiou, J. Johnson, and T. Zamenopoulos. *Embracing complexity in design*. Abingdon: Routledge.

Karimi, K. (2012). A configurational approach to analytical urban design: "Space syntax" methodology. *Urban Design International*, 17(4), pp. 297–318.

Kavouras, M. and Kokla, M. (2007). *Theories of geographic concepts: Ontological approaches to semantic integration*. Boca Raton, FL: CRC Press.

Kearns, A. and Parkinson, M. (2001). The significance of neighbourhood. *Urban Studies*, 38(12), pp. 2103–2110.

Lakoff, G. (1987). *Women, fire, and dangerous things: What categories reveal about the mind*. London: University of Chicago Press.

Latour, B. and Yaneva, A. (2008). Give me a gun and I will make all buildings move. In: R. Geiser, ed., *Explorations in architecture: Teaching, design, research*. Basel: Birkhäuser, pp. 80–89.

Lave, J. and Wenger, E. (1991). *Situated learning: Legitimate peripheral participation*. Cambridge: Cambridge University Press.

Lawton, R. (1979). Mobility in nineteenth century British cities. *The Geographical Journal*, 145(2), pp. 206–224.

Leeming, C. (2013). Why selling off homes for just £1 in a derelict area of Liverpool makes sense. *The Guardian*, 20 Feb. Available at: www.theguardian.com/commentisfree/2013/feb/20/selling-homes-liverpool [Accessed Sep. 2017].

Lefebvre, H. (1991). *The production of space*, English translation by D. Nicholson-Smith. Oxford: Blackwell Publishing.

Liverpool City Council. (2013). *Liverpool economic briefing 2013: A monitor of jobs, business and economic growth*. Liverpool: Liverpool City Council.

Liverpool City Council. (2015). *The index of multiple deprivation 2015: A Liverpool analysis*. Liverpool: Liver-pool City Council.

Liverpool City Council. (2016). *The city of Liverpool key statistics bulletin* (Issue 24, June 2016 Update). Liverpool: Liverpool City Council.

Logan, J. (2012). Making a place for space: Spatial thinking in social science. *Annual Review of Sociology*, 38, pp. 507–524.

Lupton, R. (2003). *"Neighbourhood effects": Can we measure them and does it matter?* CASE paper no. 73. Centre for Analysis of Social Exclusion, London: London School of Economics and Political Science.

Lynch, K. (1960). *The image of the city*. Cambridge, MA: MIT Press.

MacEachren, A. (2004). *How maps work: Representation, visualization, and design*. Paperback ed. London: The Guilford Press.

Marcus, L., Giusti, M. and Barthel, S. (2016). Cognitive affordances in sustainable urbanism: Contributions of space syntax and spatial cognition. *Journal of Urban Design*, 21(4), pp. 439–452. doi:10.1080/13574809.2016.1184565

Marcuse, P. (2002). The partitioned city in history. In: P. Marcuse and R. van Kempen, eds., *Of states and cities: The partitioning of urban space*. Oxford: Oxford University Press, pp. 11–34.

Massey, D. (1994). *Space, place and gender*. Cambridge: Polity Press.

McKenzie, L. (2015). *Getting by: Estates, class and culture in austerity Britain*. Bristol: Policy Press.

Medeiros, V., de Holanda, F. and Trigueiro, E. (2003). From compact colonial villages to sparse metropolis: Investigating grid integration, compactness and form of the integration core in Brazilian cities. In: J. Hanson, ed., *Proceedings of the fourth international space syntax symposium*. London: University College London.

Meyer, J. and Land, R. (2006). Threshold concepts: An introduction. In: J. Meyer and R. Land, eds., *Overcoming barriers to student understanding: Threshold concepts and troublesome knowledge*. London: Routledge.

Mezirow, J. (2000). Learning to think like an adult: Core concepts of Transformation Theory. In: J. Mezirow, ed., *Learning as transformation*. San Francisco: Jossey-Bass, pp. 3–33.

Miller, B. (1992). Collective action and rational choice: Place, community, and the limits to individual self-interest. *Economic Geography*, 68(1), *Rational Choice, Collective Action, Technological Learning*, pp. 22–42.

Morsey, H. (2012). Scarred generation. *Finance & Development*, 49(1), pp. 15–17.

Newman, O. (1996). *Creating defensible space*. Washington, DC: Office of Policy Development and Research.

Norman, D. (1988). *The design of everyday things*. New York, NY: Basic Books.

O'Brien, J. and Griffiths, S. (2017). Relating urban morphologies to movement potentials over time: A diachronic study with space syntax of Liverpool, UK. In: *Proceedings of the 11th space syntax symposium*. Lisbon, Portugal: Instituto Superior Técnico.

O'Brien, J. and Psarra, S. (2015). The dialogic city: Towards a synthesis of physical and conceptual artefacts in urban community configurations. In: K. Karimi, L. Vaughan, K. Sailer, G. Palaiologou and T. Bolton, eds., *Proceedings of the tenth international space syntax symposium*. London: University College London.

O'Brien, J., Serra, M., Hudson-Smith, A., Psarra, S., Hunter, A. and Zaltz Austwick, M. (2016). Ensuring VGI credibility in urban-community data generation: A methodological research design. *Urban Planning*, 1(2).

Palaiologou, F. and Vaughan, L. (2012). Urban rhythms: Historic housing evolution and socio-spatial boundaries. In: *Proceedings of the eighth international space syntax symposium*. Santiago, Chile.

Perlman, J. (2010). *Favela: Four decades of living on the edge in Rio de Janeiro*. Oxford: Oxford University Press.

Pinho, P. and Oliveira, V. (2009). Combining different methodological approaches to analyze the Oporto metropolitan area. In: D. Koch, L. Marcus and J. Steen, eds., *Proceedings of the Seventh International space syntax Symposium*. Stockholm, Sweden: Royal Institute of Technology (KTH).

Polovina, S. (2007). An introduction to conceptual graphs. In: *Proceedings 15th international conference on conceptual structures*. Sheffield: University of Sheffield.

Pooley, C. (1977). The residential segregation of migrant communities in mid-Victorian Liverpool. *Transactions of the Institute of British Geographers*, pp. 364–382.

Pooley, C. (2000). Patterns on the ground: Urban form, residential structure and the social construction of space. In: M. Daunton, ed., *The Cambridge urban history of Britain 1840–1950*. Cambridge: Cambridge University Press, pp. 429–465.

Pooley, C. (2016). Mobility, transport and social inclusion: Lessons from history. *Social Inclusion*, 4(3), pp. 100–109.

Power, A. (2007). *City survivors: Bringing up children in disadvantaged neighbourhoods*. Bristol: Policy Press.

Power, A. and Houghton, J. (2005). *Jigsaw cities: Big places, small spaces*. Bristol: Policy Press.

Psarra, S. (2009). The ghost of conceived space. What kind of work does or should space syntax perform for architecture. In: *Proceedings of the seventh international space syntax symposium*. Stockholm: Royal Institute of Technology.

Psarra, S., Kickert, C., and Pluviano, A. (2013). Paradigm lost: Industrial and post-industrial Detroit: An analysis of the street network and its social and economic dimensions from 1796 to the present. *Urban Design International,* 18(4), pp. 257–281.

Ramasubramanian, L. (2010). *Geographic information science and public participation.* Berlin: Spring-Verlag.

Randell, D. A., Cui, Z. and Cohn, A. G. (1992). A spatial logic based on regions and connection. In: *Proceedings 3rd international conference on knowledge representation and reasoning*. San Mateo: Morgan Kaufmann, pp. 165–176.

Rappaport, A. (1990). *The meaning of the built environment: A nonverbal communication approach.* Tucson, AZ: University of Arizona Press.

Rittel, H. and Webber, M. (1973). Dilemmas in a general theory of planning. *Policy Sciences*, 4(2), pp. 155–169.

Robinson, J. (2016). Comparative urbanism: New geographies and cultures of theorizing the Urban. *International Journal of Urban and Regional Research*, 40, pp. 187–199.

Rodger, R. and Sweet, R. (2008). The changing nature of urban history. In: *History in focus: The city*. Available at: www.history.ac.uk/ihr/Focus/City/articles/sweet.html

Roggema, R. (2017). The future of sustainable urbanism: Society-based, complexity-led, and landscape-driven. *Sustainability*, 9(8), Article 1442.

Rowe, P. (1987). *Design thinking*. Cambridge, MA: MIT Press.

RTPI. (2005). *Guidelines on effective community involvement and consultation: RTPI good practice note 1*, London: RTPI.

Sailer, K. and Penn, A. (2009). Spatiality and transpatiality in workplace environments. In: *Proceedings of the 7th international space syntax symposium*. Stockholm: KTH.

Sampson, R., Morenoff, J. and Gannon-Rowley, T. (2002). Assessing "Neighborhood Effects": Social processes and new directions in research. *Annual review of sociology*, 28, pp. 443–478.

Schön, D. (1983). *The reflective practitioner: How professionals think in action.* New York: Basic Books.

Seely-Brown, J. and Duguid, P. (1991). Organizational learning and communities of practice: Toward a unified view of working, learning, and innovation. *Organization Science*, 2(1), pp. 40–57.

Serra, M., Psarra, S. and O'Brien, J. (2018). Social and physical characterization of urban contexts: Techniques and methods for quantification, classification and purposive sampling. *Urban Planning*, 3(1), pp. 58–74.

Sibley, D. (1995). *Geographies of exclusion: Society and difference in the west*. Abingdon: Routledge.

Simon, H. (1996). *The sciences of the artificial*. 3rd ed. Cambridge, MA. MIT Press.

Smith, R. (2010). Urban studies without "scale". In: I. Farías. and T. Bender, eds., *Urban assemblages: How actor-network theory changes urban studies*. Abingdon: Routledge, Chapter 3.

SNAP. (1972). *Another chance for cities: SNAP 69|72*. Liverpool: Shelter Neighbourhood Action Project.

Sowa, J. (2008). Conceptual graphs. In: F. van Harmelen, V. Lifschitz and B. Porter, eds., *Handbook of knowledge representation*. Oxford: Elsevier, pp. 213–237.

Sykes, O., Brown, J., Cocks, M., Shaw, D. and Couch, S. (2013). A city profile of Liverpool. *Cities*, 35, pp. 299–318.

Thompson, G. (2013). *What is 'Everyday Life'? An exploration of its political connections*. CRESC Working Paper 130. University of Manchester: Centre for Research on Socio-Cultural Change (CRESC).

Tufte, E. (2006). *Beautiful evidence*. Cheshire, CT: Graphics Press.

Tulloch, D. (2007). Many, many maps: Empowerment and online participatory mapping. *First Monday*. Available at: www.firstmonday.org/ojs/index. php/fm/article/view/1620/1535

UNESCO. (2002). *Growing up in an urbanizing world*, edited by Louisa Chawla. Paris: United Nations Education, Scientific and Cultural Organization.

UN-Habitat. (2009). *Planning sustainable cities: Global report on human settlements 2009. Key findings and messages*. London: Earthscan.

UN-Habitat. (2016). *Urban planning and design lab: Brochure*. Nairobi, Kenya: United Nations Human Settlements Programme. Available at: http://unhabitat.org/urban-planning-and-design-brochure/

United Nations. (2013). *Inequality matters: Report on the world social situation 2013*. New York, NY: United Nations. Available at: www.un.org/esa/socdev/documents/reports/InequalityMatters.pdf

Urry, J. (2002). *Global complexity*. Cambridge: Blackwell Publishing.

Urry, J. (2004). The "system" of automobility. *Theory, Culture and Society*, 21(4/5), pp. 25–39.

Vanhaverbeke, W. and Cloodt, M. (2006). Open innovation in value networks. In: H. Chesborough, W. Vanhaverbeke and J. West, eds., *Open innovation: Researching a new paradigm*. Oxford: Oxford University Press.

van Steenen, M. (2010). *Graph theory and complex networks: An introduction*. Amsterdam: Maarten van Steenen.

Vaughan, L. (2007). The spatial form of poverty in Charles Booth's London. In: L. Vaughan, ed., *The spatial syntax of urban segregation*. Amsterdam: Elsevier.

Vaughan, L. (2018). *Mapping society: The spatial dimensions of social cartography*. London: UCL Press.

Vaughan, L. and Arbaci, S. (2011). The challenges of understanding urban segregation. *Built Environment*, 37(2).

Vaughan, L. and Geddes, I. (2009). Urban form and deprivation: A contemporary proxy for Charles Booth's analysis of poverty. *Radical Statistics, Social Exclusion Counted* (99).

Vaughan, L. and Penn, A. (2006). Jewish immigrant settlement patterns in Manchester and Leeds 1881. *Urban Studies*, 43(3), pp. 653–671.

von Hippel, E. (2005). *Democratizing innovation*. Cambridge, MA: MIT Press.

Walton, D. (2013). *Methods of argumentation*. Cambridge: Cambridge University Press.
Weisburd, D., Groff, E. and Yang, S. M. (2012). *The criminology of place: Street segments and our understanding of the crime problem*. Oxford: Oxford University Press.
Wenger, E. (1998). *Communities of practice: Learning, meaning, and identity*. Cambridge: Cambridge University Press.
Yaneva, A. (2012). *Mapping controversies in architecture*. Farnham: Ashgate Publishing.

Index

Note: Page numbers in *italics* indicate a figure and page numbers in **bold** indicate a table on the corresponding page.

abstract artefacts 25
actor-network theory 15, 21
affordances: design thinking and 11; movement potentials 25–26, 29, 32, 35, 38; realization of 141; social connectivity and 29; for socio-economic mobility 50; sustainable urbanism and 11–12
Alexander, C. 140
architectural theories 21
argumentation schema 105

Bender, T. 15
boundaries: movement infrastructures and 23; open space as 59; perceived 19–20, 24; road networks as 57, 79, 142; urban community spaces and 7, 9, 13, 22, 25, 99, 140
built environment 5, 11–14, 16, 22, 25

Canning Place, Central Liverpool 53, 54, *55*, 56–57
centralities 134, 137
choice 25–26
closeness graphs *131*, *133*, 134
community appraisal 28
community collaboration 3–5, 7, 12
community formations: centralities in 134; community resource affordances 108; complexity of 59; geographic barriers in 108; impact of infrastructure on 32, 35; land-use in 108, 126; North Group/South Group 106–107, *107*, 108, 114–120, 126, 129–132; open space in 118, 120; road networks and 29, 114–115, 118, 120; sentiment mapping of 118; social encounters in 126, *127–128*, 129; socio-economic contexts of 136; socio-spatial 106; spatial patterning and 26–29; urban contexts and 24, 38
community network graphs: adjusting vertex sizes 67–68; betweenness centrality 73–74; betweenness values **75**; building a graph data frame 68–70; closeness centrality 76–78; closeness network graph *77*; closeness values **76**; community weighting network graph *75*; environment for 61–62; finding asset groups 62–64; finding spatial centroids 64–65; formatting for metric observation 70–73; formatting for network analysis 73; geo-located network graph *72*; loading digital images (raster data) 66–67; North Group 129, *129*, 130, *130*; North Group closeness *131*; simple network *69*; South Group 132, *132*, *133*; South Group closeness *133*
community of practice 20
community relationships network graph: cluster diagram *91*; community assets *97*; connections between **87**; connections fixed to geo-coordinates *89*; constructing the directed graph 95–96; constructing the undirected graph 86–90; finding keyword groups 81–83; incrementing the node ranks 93–95; plot of selected *86*; processing community data 83, 85–86; programming environment 80–81; ranking the vertices 90–93; vertices weights **95**

community research: community appraisal and 28–29; community-level data in 60; emoticon symbols in 28, *28*, 29; participatory data and 79, 81; spatial patterning and 26–29
community spaces: conceptualizations of 19–20; design interventions in 14; as enacting organizations 6; impact of system-based planning on 6; intermediaries of 7; internalization of 19–20; mental models and 13, 22; movement infrastructures in 23–24; overlapping movement and 23; perceived boundaries in 19–20, 24; physical/conceptual artefacts and 12–15, 18, 21–22; policy-based interventions in 19; socio-spatial configurations of 12–18, 21–22; use of spaces in 59; weak social ties and 15; *see also* urban communities
concept graphs 100–104, *104*
conceptual artefacts: community spaces and 12–15, 18, 21–22; mental models and 13, 22, 25; non-discursive agencies and 25
crime patterns 18

Depthmap 25–26, 29, 36, 39, 56
design activities: citizen/professional networks in 14; rational problem-solving in 6–7, 99; reflective practice in 6–7, 99; reflexivity in 3, 15–16
design knowledge 6, 99–100
design practice 14, 18
design thinking 6–7, 11–13
design tools 2
dialogic city 13
dialogue 7, 20
directed graphs 134–135
disadvantaged neighbourhoods 16–17
discursive agency 14
distance 4

embodied knowledge 5
emoticon symbols 28, *28*, 29, 103, 115, 117–118, 120

Farías, I. 15
FOOTNOTE 60
Forty, A. 14
fragmentation 3

Geddes, I. 16
geo-computation 17
geographic barriers 108, *111*, *112*, 114
geographic positioning systems (GPS) 12
geographic information systems (GIS) 12, 15
Grannis, R. 17
graphic icons 103–105, 141
Groff, E. 18

Hanley, L. 19
Hanson, J. 38, 140
hierarchical design 59
Hillier, B. 14, 25, 38, 99, 140
Hippel, E. von 6
Houghton, J. 19

iGraph library 60–61, 68, 98
Index of Multiple Deprivations (IMD) 28, 33, 38, 107–108, *109*, *110*
industrialization 37–38
innovation 6, 17, 27, 37
integration 25–26, 29, 32
internal observer 38–39

knowledge representation (KR) 99–102

Lakoff, G. 102
land-use 108, *113*, 126
Latour, B. 21
Lefebvre, Henri 2–5, 140
Liverpool, UK: Canning Place 53, 54, *55*, 56–57; community case studies 106–108, 114; community formations in 26–29, *31*, 32, 35, 140; decline in 48, 50; demographic segregation in 39; deprivation in 27–28, 38, 107–108; industrialization and 37–38; LSOA data samples 33, *107*; movement potentials 26, 32–33, *34*, 35; normalized integration (NAIN) 29, 32–33, *33*; Princes Avenue 48, *49*, 50, 57; redevelopment in 50, 53, 56; regional choice evolution *44*, *45*, *46*, *47*; regional integration evolution *40*, *41*, *42*, *43*; road networks in 27, 29, 37, *37*, 39, 48, 50, 53, 56–57; Scotland Road 50, *51*, *52*, 53, 57; urban history of 36–39, 56–57; urban integration cores 29, *30*, 32, *32*; *see also* North/South Group case studies
Lower Super Output Area (LSOA) 33, *107*
Lupton, R. 19
Lynch, K. 140

Marx, Karl 2
mental models 13, 22, 25
Mezirow, J. 20
minority clustering 16
movement infrastructures 23–24
movement potentials: affordances for 25–26; Liverpool, UK 26, 32–33, *34*, 35, 48; movement infrastructures and 33, 35; natural 39; probabilistic 39; sense of place and 32; urban configurations and 36

NAIN *see* normalized integration (NAIN)
natural movement potential 39
neighbourhood effect 14–15, 19
network analysis 59
non-discursive agency 14, 18
non-human ecologies 15
normalized integration (NAIN) 29, 32–33
North/South Group case studies: common spaces in 106; community network graphs 129, *129*, 130, *130*, 131, *131*, 132, *133*, 134; conflicts in 107; deprivation in 107–108; directed graphs 134–135, *135*, 136; North Group geographic barriers *111*; North Group land use *113*; North Group spatial context *109*; open space sentiment associations *121*, *122*; road networks sentiment associations *123*, *124*; semantic associations 115–116, *116*, 117, *117*, 118, **118**, **119**, 120, 126; sense of place and 137; sentiment keywords 120, *125*; socio-economic profiles in 107–108; South Group community network *136*; South Group geographic barriers *112*; South Group land use *113*; South Group spatial context *110*; spatial assets in 120, 126, 129–132, 134–138

object artefacts 25
open access technologies 12, 60
open space 118, 120, *121*, *122*, *123*, *124*, 129

pedestrianism 12, 17
Pooley, C. 38
power 4
Power, A. 17, 19
Princes Avenue, South Liverpool 48, *49*, 50, 57
Production of Space, The (Lefebvre) 2
professionalism 4, 6
proof-of-concept 102–104

QGIS geographic information system 60

race 16–17
RCC8 icons 102–104
relational complexes 25, 140
road networks: boundaries and 57, 79, 142; community formations and 29, 114–115, 118, 120; development of 36; impact on open space 120; locked-in movement and 23; South Group *114*, *115*
R programming 39, 61, 98, 140
RStudio environment 60–61, 80

Scotland Road, North Liverpool 50, *51*, *52*, 53, 57
semantic associations: features and 118; North Group features and **118**; North Group positive/negative *116*; open space and 118, 120, 126; positive/negative 115–116; road networks and 117, 120; South Group features and **119**; South Group positive/negative *117*
sense of place 23–24, 32, 126, 137, 139–140
sentiment keywords 120, *121*, *125*
sentiment mapping 118
Sibley, D. 18
Simon, H. 140
social connectivity 29
social exclusion 17
social network research 15
socio-economic inequalities 11, 23
socio-spatial configurations 12–17
space 2–3, 21
space syntax: affordances of network structures and 141; choice in 25–26; Depthmap and 25–26; integration in 25–26; natural movements and 26, 29, 39; relational complexes and 25; spatial patterning and 23; theory of 24; topo-geometric arrangement and 20; urban configurations and 16; urban formation and 38
spatial agencies 14
spatial analysis 3, 15
spatial assets 120, 126, 129–132, 134–138
spatial complexity 12–13
spatial inequalities 11
spatial knowledge 100–102
spatial logics 102–103
spatial patterning: adaptability of 1; community formations in 26–27, *27*,

28–29; socio-economic inequalities and 23; space syntax and 23
spatial relationships: Lakoff's schema for 102, *103*; RCC8 icons 102, *102*, 103–104; representation of 102, *102*, 103–104
spatio-temporalities 38
stereotypes 18
super-positionality 13
sustainable urbanism 9–12

T-communities 17, 22
trans-spatial knowledge 100–102

urban communities: artefacts of 9, 12; challenges of 142; community formation in 24; conceptualizations of 19–20; design in 9; everyday living in 12, 23; knowledge representation (KR) in 99–100; social meanings in 13; socio-spatial configurations of 14, 16, 18; socio-spatial formations 106; spatial agencies in 14; spatial contexts of 7, 9, 23, 139–141; transformation of 141–142; *see also* community spaces
urban conceptualizations 18–20
urban configurations: crime patterns in 18; deprivation in 16; social exclusion and 17; socio-spatial 16–18; space syntax and 16; spatial logic representations 102–104; tertiary street layouts in 17
urban controversies 20–21
urban design 2–5, 7, 12
urban development 2, 16, 20–21, 36, 38, 56
urban history 38–39, 57
urban space: cognitive/historical experiences of artefacts in 100; complex artefacts and 99; controversies and 20–21; dialectics of 2–5; everyday living in 12; non-human ecologies and 15; spatial logics of 102–104; thinking of 7, 20, 99; thinking with 7, 20, 99

Vaughan, L. 16

Weisburd, D. 18
wicked problems 7

Yaneva, A. 21
Yang, S. M. 18